FEDERAL EXECUTIVE TEAM

Acting Director, Climate Change Science Program ....William J. Brennan

Acting Director, Climate Change Science Program Office ....Peter A. Schultz

Lead Agency Principal Representative to CCSP;
Division Director, Department of Energy, Office of Biological
and Environmental Research ....Jerry W. Elwood

Product Lead; Department of Energy,
Office of Biological and Environmental Research ....John C. Houghton

Synthesis and Assessment Product Advisory Group Chair,
Associate Director, EPA National Center for Environmental
Assessment...Michael W. Slimak

Synthesis and Assessment Product Coordinator,
Climate Change Science Program Office ....Fabien J.G. Laurier

OTHER AGENCY REPRESENTATIVES

Department of Agriculture ....Jan Lewandrowski
Environmental Protection Agency...Francisco de la Chesnaye
National Aeronautics and Space Administration ....Phil DeCola
National Oceanic and Atmospheric Administration ....Hiram Levy
National Oceanic and Atmospheric Administration ....Larry W. Horowitz

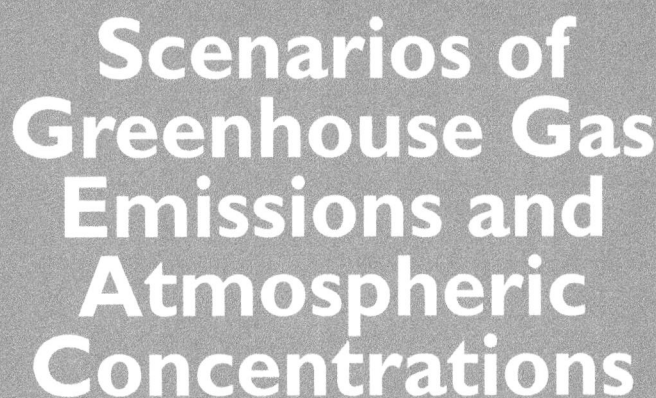

Scenarios of Greenhouse Gas Emissions and Atmospheric Concentrations

Synthesis and Assessment Product 2.1a
Report by the U.S. Climate Change Science Program
and the Subcommittee on Global Change Research

AUTHORS:

Leon E. Clarke, Pacific Northwest National Laboratory
James A. Edmonds, Pacific Northwest National Laboratory
Henry D. Jacoby, Massachusetts Institute of Technology
Hugh M. Pitcher, Pacific Northwest National Laboratory
John M. Reilly, Massachusetts Institute of Technology
Richard G. Richels, Electric Power Research Institute

CLIMATE CHANGE SCIENCE PROGRAM PRODUCT DEVELOPMENT ADVISORY COMMITTEE

Six members of the Climate Change Science Program Product Development Advisory Committee (CPDAC) wrote this Climate Change Science Program Synthesis and Assessment Product at the request of the Department of Energy. The entire CPDAC has accepted the contents of the product.

Chair
Robert M. White, The Washington Advisory Group

Vice Chair
Soroosh Sorooshian, University of California-Irvine

Designated Federal Officer
Anjuli S. Bamzai, Department of Energy
Office of Biological and Environmental Research

Members

David C. Bader
Lawrence Livermore National
Laboratory

Virginia R. Burkett
U.S. Geological Survey

Antonio J. Busalacchi
University of Maryland

*Leon E. Clarke
Pacific Northwest National
Laboratory

Curtis C. Covey
Lawrence Livermore National
Laboratory

*James A. Edmonds
Pacific Northwest National
Laboratory

Karen Fisher-Vanden
Dartmouth College

Brian P. Flannery
Exxon-Mobil Corporation

William J. Gutowski
Iowa State University

David G. Hawkins
Natural Resources Defense Council

Isaac M. Held
Geophysical Fluid Dynamics
Laboratory

*Henry D. Jacoby
Massachusetts Institute of
Technology

David W. Keith
University of Calgary

Kenneth E. Kunkel
Illinois State Water Survey

Richard S. Lindzen
Massachusetts Institute of
Technology

Linda O. Mearns
National Center for Atmospheric
Research

Ronald L. Miller
National Aeronautics and Space
Administration

Edward A. Parson
University of Michigan

*Hugh M. Pitcher
Pacific Northwest National
Laboratory

William A. Pizer
Resources for the Future

*John M. Reilly
Massachusetts Institute of
Technology

*Richard G. Richels
Electric Power Research Institute

Cynthia E. Rosenzweig
National Aeronautics and Space
Administration

Robin T. Tokmakian
Naval Postgraduate School

Mort D. Webster
Massachusetts Institute of
Technology

Julie A. Winkler
Michigan State University

Gary W. Yohe
Wesleyan University

Minghua H. Zhang
Stony Brook University

*Authors for *Scenarios of Greenhouse Gas Emissions and Atmospheric Concentrations*

July 2007

Members of Congress:

On behalf of the National Science and Technology Council, the U.S. Climate Change Science Program (CCSP) is pleased to transmit to the President and the Congress this report, *Scenarios of Greenhouse Gas Emissions and Atmospheric Concentrations and Review of Integrated Scenario Development and Application.* This is the second in a series of Synthesis and Assessment Products produced by the CCSP. This series of 21 reports is aimed at providing current evaluations of climate change science to inform public debate, policy, and operational decisions. These reports are also intended to help inform CCSP's consideration of future program priorities. This second Synthesis and Assessment Product issued pursuant to Section 106 of the Global Change Research Act of 1990 and has two components: "Development of New Scenarios of Greenhouse Gas Emissions and Atmospheric Concentrations" (Part A) and a "Review of Integrated Scenario Development and Application" (Part B). Here we transmit to you Part A.

CCSP's guiding vision is to provide the Nation and the global community with the science-based knowledge to manage the risks and opportunities of change in the climate and related environmental systems. The Synthesis and Assessment Products are important steps toward that vision, helping translate CCSP's extensive observational and research base into informational tools directly addressing key questions that are being asked of the research community.

This product will contribute to and enhance the ongoing and iterative international process of producing and refining climate-related scenarios and scenario tools. It was developed with broad scientific input and in accordance with the Guidelines for Producing CCSP Synthesis and Assessment Products, Section 515 of the Treasury and General Government Appropriations Act for Fiscal Year 2001 (Public Law 106-554), and the Information Quality Act guidelines issued by the Department of Energy pursuant to Section 515. The CCSP Interagency Committee relies on Department of Energy certifications regarding compliance with Section 515 and the Guidelines for Producing CCSP Synthesis and Assessment Products.

We commend the report's authors for both the thorough nature of their work and their adherence to an inclusive review process.

Samuel W. Bodman
Secretary of Energy

Chair, Committee on
Climate Change
Science and Technology Integration

Carlos M. Gutierrez
Secretary of Commerce

Vice-Chair, Committee on
Climate Change
Science and Technology Integration

John H. Marburger III, Ph.D.
Director, Office of
Science and Technology Policy
Executive Director, Committee on
Climate Change
Science and Technology Integration

ACKNOWLEDGEMENT

This report has been peer reviewed in draft form by individuals chosen for their diverse perspectives and technical expertise. The expert review and selection of reviewers followed the OMB's Information Quality Bulletin for Peer Review. The purpose of this independent review is to provide candid and critical comments that will assist the Climate Change Science Program in making this published report as sound as possible and to ensure that the report meets institutional standards. The peer review comments, draft manuscript, and response to the peer review comments are publicly available at: www.climatescience.gov/Library/sap/sap2-1/default.php.

We wish to thank the following individuals for their peer review of this report:

Joseph Aldy, Resources for the Future

Bill Chameides, James Wang, Environmental Defense

Russell Jones, American Petroleum Institute

David Rind, NASA Goddard Institute for Space Studies

Brent Sohngen, Ohio State University

Richard Tol, Hamburg, Vrije and Carnegie Mellon Universities

John Weyant, Stanford University

We would also like to thank the numerous individuals who provided their comments during the public comment period. The public review comments, draft manuscript, and response to the public comments are publicly available at: www.climatescience.gov/Library/sap/sap2-1/default.php.

The authors also wish to acknowledge the contributions of Geoffrey Blanford, Josh Lurz, Sergey Paltsev, Thomas Rutherford, and Marshall Wise in developing these scenarios.

EDITORIAL TEAM

Editor..Loel Kathmann

Technical Advisor ....David Dokken

Graphic Production ....DesignConcept

Recommended Citations

The Entire Volume - CCSP Synthesis and Assessment Product 2.1
CCSP, 2007: Scenarios of Greenhouse Gas Emissions and Atmospheric Concentrations (Part A) and Review of Integrated Scenario Development and Application (Part B). A Report by the U.S. Climate Change Science Program and the Subcommittee on Global Change Research [Clarke, L., J. Edmonds, J. Jacoby, H. Pitcher, J. Reilly, R. Richels, E. Parson, V. Burkett, K. Fisher-Vanden, D. Keith, L. Mearns, C. Rosenzweig, M. Webster (Authors)]. Department of Energy, Office of Biological & Environmental Research, Washington, DC., USA, 260 pp.

This Sub-Report (2.1A)
Clarke, L., J. Edmonds, H. Jacoby, H. Pitcher, J. Reilly, R. Richels, 2007. *Scenarios of Greenhouse Gas Emissions and Atmospheric Concentrations*. Sub-report 2.1A of Synthesis and Assessment Product 2.1 by the U.S. Climate Change Science Program and the Subcommittee on Global Change Research. Department of Energy, Office of Biological & Environmental Research, Washington, DC., USA, 154 pp.

The Companion Sub-Report (2.1B)
Parson, E., V. Burkett, K. Fisher-Vanden, D. Keith, L. Mearns, H. Pitcher, C. Rosenzweig, M. Webster, 2007. *Global Change Scenarios: Their Development and Use*. Sub-report 2.1B of Synthesis and Assessment Product 2.1 by the U.S. Climate Change Science Program and the Subcommittee on Global Change Research. Department of Energy, Office of Biological & Environmental Research, Washington, DC., USA, 106 pp.

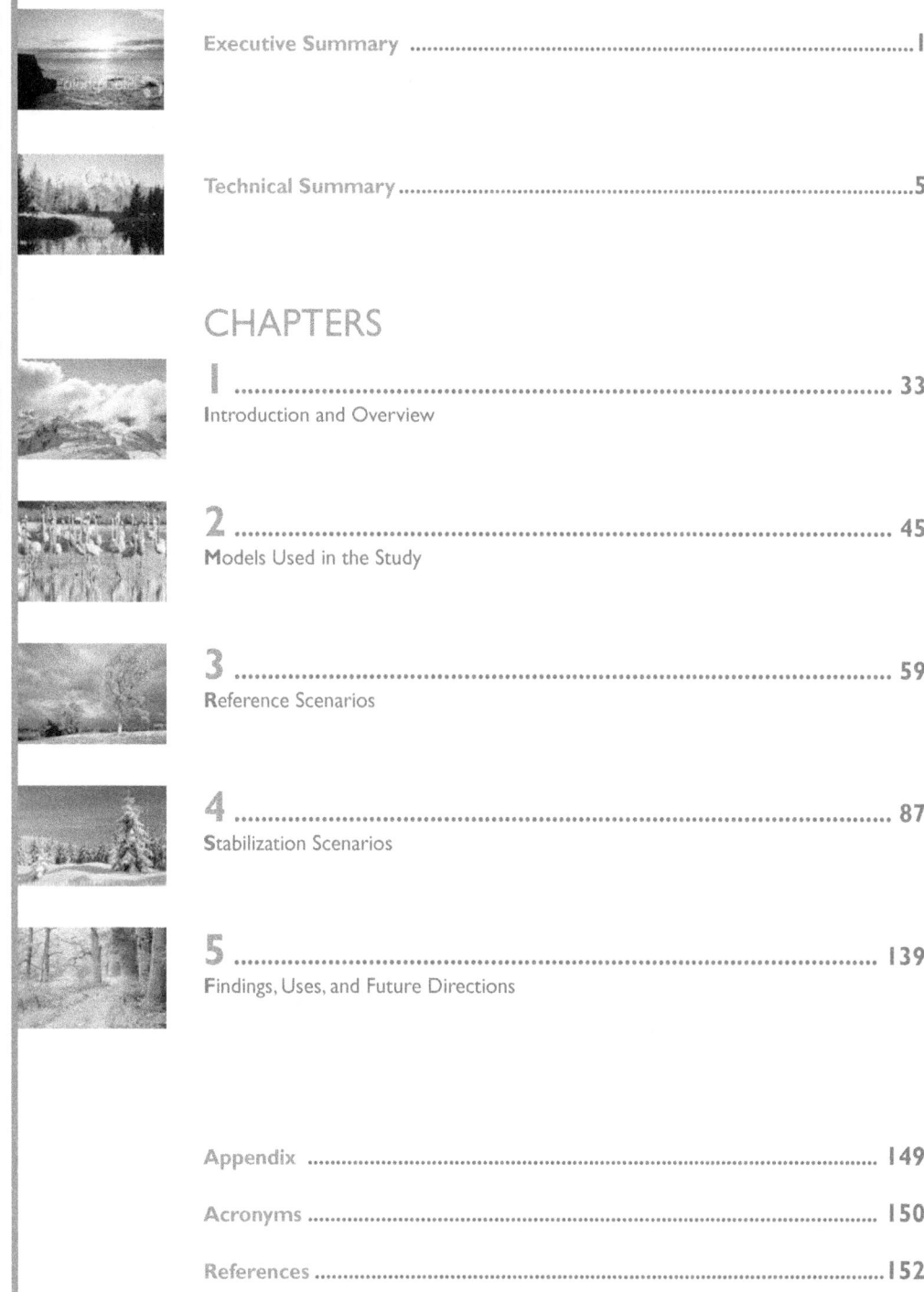

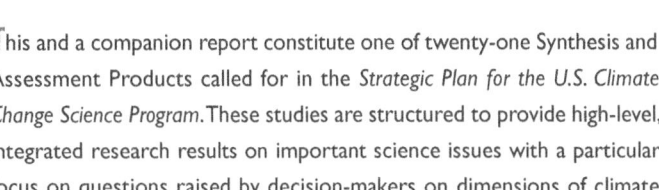

Executive Summary

INTRODUCTION

This and a companion report constitute one of twenty-one Synthesis and Assessment Products called for in the *Strategic Plan for the U.S. Climate Change Science Program*. These studies are structured to provide high-level, integrated research results on important science issues with a particular focus on questions raised by decision-makers on dimensions of climate change directly relevant to the U.S. One element of the CCSP's strategic vision is to provide *decision support tools* for differentiating and evaluating response strategies. Scenario-based analysis is one such tool. The scenarios in this report explore the implications of alternative stabilization levels of anthropogenic greenhouse gases (GHGs) in the atmosphere, and they explicitly consider the economic and technological foundations of such response options. Such scenarios are a valuable complement to other scientific research contained in the twenty-one CCSP Synthesis and Assessment Products. The companion to the research reported here, *Global-Change Scenarios: Their Development and Use*, explores the broader strategic frame for developing and utilizing scenarios in support of climate decision making.

STUDY DESIGN

The scenarios in this report were developed using integrated assessment models (IAMs). These analysis capabilities integrate computer models of socioeconomic and technological determinants of the emissions of GHGs with models of the natural science of Earth system response, including the atmosphere, oceans, and terrestrial biosphere. Three IAMs were applied in the scenario development:

- The Integrated Global Systems Model (IGSM) of the Massachusetts Institute of Technology's Joint Program on the Science and Policy of Global Change.

- The Model for Evaluating the Regional and Global Effects (MERGE) of GHG reduction policies developed jointly at Stanford University and the Electric Power Research Institute.

- The MiniCAM Model of the Joint Global Change Research Institute, a partnership between the Pacific Northwest National Laboratory and the University of Maryland.

Each modeling group first produced a reference scenario under the assumption that no climate policies are imposed beyond current commitments, namely the 2008-12 first period of the Kyoto Protocol and the U.S. goal of reducing reduce GHG emissions per unit of its gross domestic product by 18% by 2012. The resulting reference cases are not predictions or best-judgment forecasts but scenarios designed to provide clearly defined points of departure for studying the implications of alternative stabilization goals. As instructed in the Prospectus for the study, the modeling teams used model input assumptions they considered *meaningful* and *plausible*. The resulting reference scenarios provide insights into how the world might evolve <u>without</u> additional efforts to constrain GHG emissions, given various assumptions about principal drivers of these emissions such as population increase, economic growth, land and labor productivity growth, technological options, and resource endowments.

Each modeling group then produced four additional stabilization scenarios framed as departures from its reference scenario. The stabilization levels are common across the modeling groups and are defined in terms of the total long-term effect on the Earth's heat balance of the combined effect of the primary anthropogenic GHGs: carbon dioxide (CO_2), nitrous oxide (N_2O), methane (CH_4), hydrofluorocarbons (HFCs), perfluorocarbons (PFCs), and sulfur hexafluoride (SF_6). The potential for climate-related controls on other human emissions, such as aerosols and their precursors, was not incorporated into the stabilization constraints, although the participating models represent the emissions of many of these substances. With the exception of these stabilization levels, and a common hypothesis about the sharing among nations of the mitigation task, there was no direct coordination among the modeling groups either in the assumptions underlying the reference scenario or the precise paths to stabilization.

The results drawn from the simulations were selected to provide insight into questions such as the following:

- What emissions trajectories over time are consistent with meeting the four stabilization levels, and what are the key factors that shape them?

- What energy system characteristics are consistent with each of the four alternative stabilization levels, and how might these characteristics differ among stabilization levels?

- What are the possible economic consequences of meeting each of the four alternative stabilization levels?

With its focus on reducing emissions to meet various stabilization levels the study does not explore climate damages that might be avoided or ancillary benefits (such as lower air pollution) of emissions reduction. Thus, though the scenarios provide a useful input to climate-related decision making they address only one of several components of a cost-benefit analysis of climate policy. In addition , although these scenarios incorporate new thinking on GHG emissions and possible mitigation paths they were not designed to span the full range of possible futures or to provide an uncertainty analysis of key forces. They are intended, rather, to enhance understanding of the implications of different ways that the future might evolve without assigning likelihoods to outcomes.

POTENTIAL APPLICATIONS

There are many potential applications of scenarios of this form, and to facilitate their use the numerical results are provided in a companion data set. Possible users include climate modelers and the science community; those involved in national public policy formulation; managers of Federal research programs; state and local government officials who face decisions that might be affected by climate change and mitigation measures; and individual firms, non-governmental organizations, and members of the public. Insights from the scenarios may be used directly as inputs to the decision-making processes, or the scenarios may serve as inputs to further analyses in support of climate decision making. A sample of possible further analyses would include the following:

- The scenarios can provide a basis for study of the climate implications of alternative stabilization levels, as an input to climate models, and then to follow-on studies of potential climate impacts.

- The scenarios can serve as a point of departure for exploring possible technology cost and performance goals, using information from the scenarios on energy prices and technology deployment levels.

- The scenarios can provide a foundation for analysis of the non-climate environmental implications of new energy sources at large scale.

- The scenarios could serve as an input to a more complete analysis of the economic effects of stablizing and the different radiative forcing levels, such as indicators of consumer impact in the U.S.

- The scenarios can be applied in comparative mode, extending the lessons to be learned from the three models in this research to those to be gained from scenarios developed using different approaches.

The varied clientele for these scenarios and the variety of questions they might inform implies a highly diverse set of possible needs, and no single scenario exercise can hope to fully satisfy all of them. Therefore these scenarios likely will stimulate further questions and the demand for more detailed analysis, some of which might be satisfied by further scenario development from models like those used here, but others demanding detail that can only be provided with alternative modeling and analysis techniques.

Several characteristics of these scenarios make them particularly valuable for these and other types of applications. One advantage is the update of economic and technology data and assumptions and the use of improved scenario development tools. It has been over a decade since the last emissions scenario development project of the Intergovernmental Panel on Climate Change (IPCC) – its *Special Report on Emissions Scenarios* (SRES) – and over five years since the subsequent CO_2 stabilization scenarios in the IPCC's Third Assessment Report. Over this time, substantial advances have been made in both economic and natural science components of the IAMs used to simulate the various scenarios. A second advance of this research is its all-gas approach. Many other stabilization scenarios have focused on CO_2 with little attention to other human influences. The scenarios presented here consider stabilization in terms of the combined effect of all six categories of GHGs listed earlier so that the full range of policy options is considered simultaneously. Finally, there is great advantage in the simultaneous application to the task, and parallel presentation of results, by three independent modeling groups, applying IAMs each of which has its own special strengths. Comparison of scenarios across the models provides useful insights into the role of key assumptions, the realms of most fruitful technology development, and aspects of the natural science (particularly the carbon cycle) that have a substantial effect on the difficulty of the stabilization task.

SCENARIO HIGHLIGHTS

The report and supporting database provide many details of the implications for the U.S. and global economy, with particular focus on the energy sector, for the reference conditions and the four levels of possible atmospheric stabilization. Highlights of the picture that is found there include the following:

In the reference scenarios, economic and energy growth, combined with continued fossil fuel use, lead to changes in the Earth's radiation balance that are three to four times that already experienced since the beginning of the industrial age. By 2100, primary energy consumption increases from over 3 to nearly 4 times 2000 levels as economic growth outpaces improvements in the efficiency of energy use. Non-fossil energy use grows from over 4 to almost 9 times over the century, but this growth is insufficient to supplant fossil fuels as the major source of energy. As a result, global CO_2 emissions more than triple between 2000 and 2100, and emissions are rising at the end of the twenty-first century in all three reference scenarios. Combined with the effects of non-CO_2 GHGs, the increase in anthropogenic radiative forcing from preindustrial levels is substantial.

In the stabilization scenarios, CO_2 emissions peak and decline during the twenty-first century or soon thereafter. Emissions of non-CO_2 GHGs are also reduced. The timing of GHG emissions reductions varies substantially across the four radiative forcing stabilization levels. Under the most stringent stabilization levels, CO_2 emissions begin to decline immediately or within a matter of decades. Under the less stringent stabilization levels, CO_2 emissions do not peak until late in the century or beyond, and they are 1½ to over 2½ times today's levels in 2100.

In the stabilization scenarios, GHG emissions reductions require a transformation of the global energy system, including reductions in the demand for energy (relative to the reference scenarios) and changes in the mix of energy technologies and fuels. This transformation is more substantial and takes place more quickly at the more stringent stabilization levels. Fossil fuel use and energy consumption are reduced in all the stabilization scenarios due to increased

consumer prices for fossil fuels. Use of shale oil, tar sands, and synthetic fuels from coal are greatly reduced or, under the most stringent stabilization levels, eliminated. Across the stabilization scenarios, CO_2 emissions from electric power generation are reduced at relatively lower prices than CO_2 emissions from other sectors, such as transport, industry, and buildings. Emissions are reduced from electric power by increased use of technologies such as CO_2 capture and storage, nuclear energy, and renewable energy. Other sectors respond to rising GHG prices by reducing demands for fossil fuels; substituting low- or non-emitting energy sources such as bioenergy and low-carbon electricity or hydrogen; and applying CO_2 capture and storage where possible.

Substantial differences in GHG emissions prices and associated economic costs arise among the modeling groups for each stabilization level. These differences are illustrative of some of the unavoidable uncertainties in long-term scenarios. Among the most important factors influencing the variation in economic costs are: (1) differences in assumptions – such as those regarding economic growth over the century, the behavior of the oceans and terrestrial biosphere in taking up CO_2, and opportunities for reduction in non-CO_2 GHG emissions – that determine the amount that CO_2 emissions that must be reduced to meet the radiative forcing stabilization levels; and (2) differences in assumptions about technologies, particularly in the second half of the century, to shift final demand to low-carbon sources such as biofuels and low-carbon electricity or hydrogen, in transportation, industrial, and buildings end uses. All other things being equal, scenarios with more low-cost technology options and lower required emissions reductions have lower economic costs.

FOLLOW-ON EFFORTS

Generating scenarios is not a once-and-for-all activity. The scenarios in this report represent but one step in a long process of research and assessment, continuing an over 20-year tradition of research and analysis in the climate area. They will need to be updated as knowledge advances and conditions change. Indeed, the research presented here suggests several areas of potentially fruitful research:

- Analysis of the sensitivity of results to assumptions about the cost, performance and environmental issues surrounding key technologies such as nuclear power, carbon capture and storage, and biofuels.

- Consideration of scenarios based on different assumptions than used here about the way that the burdens of emissions mitigation may be shared among nations and over time.

- Expansion and improvement of analysis of human land use and the terrestrial carbon cycle.

- Inclusion of other anthropogenic emissions that affect the Earth's heat balance, such as the different types of aerosols, and the effect of the tropospheric ozone (another GHG) that results from urban air pollution.

- Addition of uncertainty analysis and consideration of decision-making under these conditions.

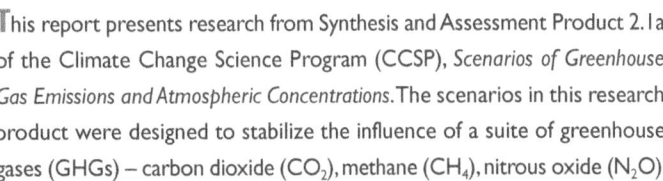

Technical Summary

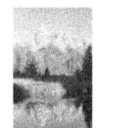

HIGHLIGHTS OF THE REPORT

Background

This report presents research from Synthesis and Assessment Product 2.1a of the Climate Change Science Program (CCSP), *Scenarios of Greenhouse Gas Emissions and Atmospheric Concentrations*. The scenarios in this research product were designed to stabilize the influence of a suite of greenhouse gases (GHGs) – carbon dioxide (CO_2), methane (CH_4), nitrous oxide (N_2O), hydrofluorocarbons (HFCs), perfluorocarbons (PFCs) and sulfur hexafluoride (SF_6) – on the Earth's radiation balance, measured in terms of radiative forcing. Four radiative forcing stabilization levels are considered. The resulting atmospheric concentrations of the largest single contributor, CO_2, are roughly 450, 550, 650 and 750 parts per million by volume (ppmv). Responding to the Prospectus for this research product (CCSP 2005), this report focuses on (1) GHG emissions trajectories, (2) global and U.S. energy system implications, and (3) economic implications of stabilization.

This research was conducted using computer-based research tools known as integrated assessment models. Three modeling groups each independently developed a reference scenario, in which all climate policies were assumed to expire in 2012, and then developed four stabilization scenarios as departures from their respective reference scenarios. Idealized emissions-reduction measures – designed to achieve emissions reductions wherever, whenever, and using whichever GHG was most cost effective – were imposed to limit GHG emissions and meet the four radiative forcing stabilization levels. Evidence from previous literature suggests that if less idealized measures were employed to stabilize radiative forcing, the costs could be substantially higher. Further, this research considers only the costs of stabilization; it does not consider the benefits of potential climate change avoided or of possible ancillary benefits of emissions reduction, such as reduced air pollution.

The scenarios in this report are not predictions or best-judgment forecasts from the modeling groups. Rather, they constitute new research intended to advance understanding of the forces that lead to GHG emissions and that shape opportunities to stabilize GHG concentrations and radiative forcing. Although the future is uncertain and the scenarios are strongly dependent on many underlying assumptions, this research provides useful insights for those engaged in climate-related decision making.

Highlights of the Report

In the reference scenarios, economic and energy growth, combined with continued fossil fuel use, lead to changes in the Earth's radiation balance that are three to four times that already experienced since the beginning of the industrial age. By 2100, primary energy consumption increases from over three to nearly four times 2000 levels as economic growth outpaces improvements in the efficiency of energy use. Non-fossil energy use grows from over four to almost nine times over the century, but this growth is insufficient to supplant fossil fuels as the major source of energy. As a result, global CO_2 emissions more than triple between 2000 and 2100, and emissions are rising at the end of the twenty-first century in all three reference scenarios. Combined with the effects of non-CO_2 GHGs, the increase in anthropogenic radiative forcing from preindustrial levels is substantial.

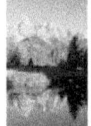

In the stabilization scenarios, CO_2 emissions peak and decline during the twenty-first century or soon thereafter. Emissions of non-CO_2 GHGs are also reduced. The timing of GHG emissions reductions varies substantially across the four radiative forcing stabilization levels. Under the most stringent stabilization levels, CO_2 emissions begin to decline immediately or within a matter of decades. Under the less stringent stabilization levels, CO_2 emissions do not peak until late in the century or beyond, and they are 1½ to over 2½ times today's levels in 2100.

In the stabilization scenarios, GHG emissions reductions require a transformation of the global energy system, including reductions in the demand for energy (relative to the reference scenarios) and changes in the mix of energy technologies and fuels. This transformation is more substantial and takes place more quickly at the more stringent stabilization levels. Fossil fuel use and energy consumption are reduced in all the stabilization scenarios due to increased consumer prices for fossil fuels. Use of shale oil, tar sands, and synthetic fuels from coal are greatly reduced or, under the most stringent stabilization levels, eliminated. Across the stabilization scenarios, CO_2 emissions from electric power generation are reduced at relatively lower prices than CO_2 emissions from other sectors, such as transport, industry, and buildings. Emissions are reduced from electric power by increased use of technologies such as CO_2 capture and storage (CCS), nuclear energy, and renewable energy. Other sectors respond to rising greenhouse gas prices by reducing demands for fossil fuels; substituting low- or non-emitting energy sources such as bioenergy and low-carbon electricity or hydrogen; and applying CCS where possible.

Substantial differences in GHG emissions prices and associated economic costs arise among the modeling groups for each stabilization level. These differences are illustrative of some of the unavoidable uncertainties in long-term scenarios. Among the most important factors influencing the variation in economic costs are: (1) differences in assumptions – such as those regarding economic growth over the century, the behavior of the oceans and terrestrial biosphere in taking up CO_2, and opportunities

for reduction in non-CO_2 GHG emissions – that determine the amount that CO_2 emissions must be reduced to meet the radiative forcing stabilization levels; and (2) differences in assumptions about technologies, particularly in the second half of the century, to shift final demand to low-carbon sources such as biofuels and low-carbon electricity or hydrogen, in transportation, industrial, and buildings end uses. All other things being equal, scenarios with more low-cost technology options and lower required emissions reductions have lower economic costs.

BACKGROUND

The Strategic Plan for the U.S. Climate Change Science Program (CCSP 2003) noted that "…sound, comprehensive emissions scenarios are essential for comparative analysis of how climate might change in the future, as well as for analyses of mitigation and adaptation options." The Plan includes Product 2.1, *Scenarios of Greenhouse Gas Emissions and Atmospheric Concentrations and Review of Integrated Scenario Development and Application,* which consists of two parts. This report presents the scenario development component (Product 2.1A); the review of scenario methods (Product 2.1B) is the subject of a separate report (CCSP 2007).

Guidelines for producing these scenarios were set forth in a Prospectus, which specified that the new scenarios focus on alternative levels of atmospheric stabilization of the radiative forcing from the combined effects of a suite of the main anthropogenic GHGs. The Prospectus also set forth criteria for the facilities to be used in the analysis. Scenarios developed using three models that meet the Prospectus conditions are reported here.

The scenarios in this report are intended as one of many inputs to public and private discussions regarding climate change and what to do about it, and they may serve as a point of departure for further CCSP and other analyses that might inform these discussions in the future. The possible users of these scenarios are many and diverse. They include climate modelers and the science community; those involved in national public policy formulation; managers of Federal research programs; state and local government officials who face decisions that might be affected by climate change and mitigation measures; and individual firms, non-governmental organizations, and members of the public. Such a varied clientele implies an equally diverse set of possible needs, and no single scenario exercise can hope to fully satisfy all of these needs.

Each of the three modeling groups participating in this research first developed a no-climate policy scenario – referred to as a reference scenario – which serves as baseline for development of alternative scenarios with emissions control. Each modeling group then developed four control scenarios leading to stabilization of radiative forcing at four alternative levels. The resulting scenarios provide insight into questions such as the following:

- What emissions trajectories over time are consistent with meeting the four alternative stabilization levels, and what are the key factors that shape them?

- What energy system characteristics are consistent with each of the four alternative stabilization levels, and how might these characteristics differ among stabilization levels?

- What are the possible economic consequences of meeting each of the four alternative stabilization levels?

Although each of the models used to develop these scenarios represents the world as a set of interconnected nations and multi-nation regions, as specified in the Prospectus, this report focuses on the U.S. and world characteristics of the scenarios.

With the exception of the stabilization levels themselves and a common hypothesis about international burden sharing, there was no direct coordination among the modeling groups either in the assumptions underlying the reference scenario or the precise path to stabilization. Furthermore, the scenarios were not designed to span the full range of possible futures, and no explicit uncertainty analysis was called for. Although the future is uncertain and the scenarios depend on many underlying assumptions, this research illuminates a range of possible future

developments and provides useful insights for those engaged in climate-related decision making.

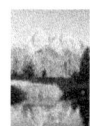

The scenarios in this report do not constitute a cost-benefit analysis of climate policy. They focus exclusively on the issues associated with reducing emissions to meet various stabilization levels; they do not consider the damages avoided through stabilization or ancillary benefits that could be realized by emissions reductions, such as reductions in local air pollution. Thus, although the scenarios should serve as a useful input to climate-related decision making, they address only one of several components of a benefit-cost analysis of climate policy.

Scenario research such as this continues a tradition of research and analysis that has gone on for over 20 years. This work will be continued and refined as the field advances, new information becomes available, and decision makers raise new questions and issues. Similar work is conducted by modeling groups in Europe and Asia. The scenarios developed here add to this larger body of scholarship and should be viewed as one additional piece of information in an ongoing and iterative process of scenario development.

MODELS USED TO DEVELOP THE SCENARIOS

The Prospectus for this research set out the following criteria for participating models: they must (1) be global in scale, (2) be capable of producing global emissions totals for designated GHGs, (3) represent multiple regions, (4) be capable of simulating the radiative forcing from these GHGs and substances, (5) have technological resolution capable of distinguishing among major sources of primary energy (e.g., renewable energy, nuclear energy, biomass, oil, coal, and natural gas) as well as between fossil fuel technologies with and without carbon capture and storage systems, (6) be economics-based and capable of simulating macroeconomic cost implications of stabilization, and (7) look forward at least to the end of the twenty-first century. In addition, modeling groups were required to have a track record of publications in professional, refereed journals, specifically in the use of their models for the analysis of long-term GHG emission scenarios.

Application of these criteria led to the selection of three models:

- The Integrated Global Systems Model (IGSM) of the Massachusetts Institute of Technology's Joint Program on the Science and Policy of Global Change

- The Model for Evaluating the Regional and Global Effects (MERGE) of GHG reduction policies developed jointly at Stanford University and the Electric Power Research Institute.

- The MiniCAM Model of the Joint Global Change Research Institute, a partnership between the Pacific Northwest National Laboratory and the University of Maryland.

Each of these models has been used extensively for climate change analysis. The roots of each extend back more than a decade, during which time features and details have been refined, modified, and added. Research using each has appeared widely in peer-reviewed publications.

APPROACH

As directed by the Prospectus, each of the three modeling groups produced one reference scenario and four stabilization scenarios, for a total of 15 scenarios. First, the reference scenarios were developed under the assumption that no climate policy would be implemented beyond the set of policies currently in place (e.g., the

Kyoto Protocol and the U.S. carbon intensity goal, each terminating in 2012 because goals beyond that date have not been identified). Each modeling group developed its own reference scenario. The Prospectus required only that each reference scenario be based on assumptions believed by the participating modeling groups to be *meaningful* and *plausible*. Each of the three reference scenarios is based on a different set of assumptions about how the future might unfold without additional climate policies. These assumptions are not intended as predictions or best-judgment forecasts of the future by the respective modeling groups. Rather, they represent possible paths that the future might follow to serve as a platform for examining how emissions might be reduced to achieve stabilization.

Each group then produced four stabilization scenarios by constraining the models to achieve four alternative radiative forcing levels. Stabilization was defined in terms of the total long-term radiative impact of a suite of GHGs including CO_2, N_2O, CH_4, HFCs, PFCs, and SF_6. These are the gases enumerated in the U.S. goal to reduce the intensity of GHG emissions relative to gross domestic product (GDP) as well as the Kyoto Protocol. Other substances with radiative impact, such as gases controlled under the Montreal Protocol, carbon monoxide (CO), ozone (O_3), and aerosols were not included in the radiative forcing levels.

The four radiative forcing stabilization scenarios were developed so that the combined radiative forcing from these GHGs since preindustrial times was constrained to no more than 3.4 W/m^2 for Level 1, 4.7 W/m^2 for Level 2, 5.8 W/m^2 for Level 3, and 6.7 W/m^2 for Level 4. Because radiative forcing was defined relative to preindustrial times, it includes the

Table TS.1. Greenhouse Gas Concentrations and Forcing.
Concentrations of GHGs have increased since 1750 (preindustrial), altering the radiative energy budget of the Earth's climate system.

	Preindustrial Concentration (1750)	Current Concentration (1998)	Contribution to Radiative Forcing, (W/m^2, 1750 to 1998)
CO_2	278 ppmv	365 ppmv	1.46
CH4	700 ppbv	1745 ppbv	0.48
N_2O	270 ppbv	314 ppbv	0.15
HFCs, PFCs, SF_6	0	various	$\approx$ 0.02
Total	—	—	$\approx$ 2.1
Source: IPCC 2001.			

	Total Radiative Forcing from GHGs in this Research (W/m²)	Approximate Contribution to Radiative Forcing from non-CO₂ GHGs (W/m²)	Approximate Contribution to Radiative Forcing from CO₂ (W/m²)	Corresponding CO₂ Concentration (ppmv)
Level 1	3.4	0.8	2.6	450
Level 2	4.7	1.0	3.7	550
Level 3	5.8	1.3	4.5	650
Level 4	6.7	1.4	5.3	750
Year 1998	≈ 2.1	0.65	1.46	365
Preindustrial (1750)	—	—	—	278

Table TS.2. Radiative Forcing Stabilization Levels (W/m² from preindustrial) and Approximate Resulting CO₂ Concentrations (ppmv). The radiative forcing levels were constructed so that the CO_2 concentrations resulting from stabilization of total radiative forcing, after accounting for radiative forcing from the non-CO_2 GHGs, would be roughly 450 ppmv, 550 ppmv, 650 ppmv, and 750 ppmv.

roughly 2.1 W/m² of radiative forcing from these substances that had already occurred from 1750 to 1998 (Table TS.1).

These radiative forcing stabilization levels were chosen so that the associated CO_2 concentrations would be roughly 450 ppmv, 550 ppmv, 650 ppmv, and 750 ppmv after accounting for the contributions to radiative forcing from the non-CO_2 GHGs (Table TS.2). If these CO_2 concentrations were achieved exactly, the radiative forcing from CO_2 would be less than the radiative forcing stabilization levels because of the allowance for additional forcing from the non-CO_2 GHGs. Thus, the radiative forcing stabilization levels should not be interpreted as the "CO_2–equivalent" levels associated with the approximate CO_2 concentrations in Table TS.2. Because the stabilization exercises sought least-cost reductions among the gases, any correspondence between radiative forcing levels and CO_2 concentrations is necessarily approximate and differs among modeling groups because of differences in the treatment of the forces that influence emissions of GHGs, possibilities for emissions reductions, and tradeoffs between reductions among GHGs.

OVERVIEW OF THE SCENARIOS

This section provides an overview of the scenarios. The three reference scenarios are discussed in the next section, followed by a discussion of the twelve stabilization scenarios, four from each modeling group.

Reference Scenarios

The difficulty of achieving any specified level of atmospheric stabilization depends heavily on the emissions that would occur absent actions to address GHG emissions. In other words, the reference scenario strongly influences the stabilization scenarios. If the reference scenario has inexpensive fossil fuels and high-economic growth, then larger changes to the energy sector and other parts of the economy may be required to stabilize radiative forcing. On the other hand, if the reference scenario shows lower economic growth and emissions, and perhaps increased exploitation of non-fossil sources even in the absence of climate policy, then the effort required to stabilize radiative forcing will not be as great.

Energy production, transformation, and consumption are central features in all of these scenarios, although non-CO_2 gases and changes in land use also make a significant contribution to aggregate GHG emissions. Demand for energy over the coming century will be driven by economic growth and will also be strongly influenced by the way that energy systems respond to depletion of resources, changes in prices, and improvements in technology. Demand for energy in developed countries remains strong in all the scenarios and is even stronger in developing countries, where millions of people seek greater access to commercial energy. These developments strongly influence the emissions of GHGs, their disposition, and the resulting change in radiative forcing in the reference scenarios.

Figure TS.1. Global Primary Energy Consumption Across Reference Scenarios (EJ/yr). Global primary energy consumption rises in all three reference scenarios, from about 400 EJ/yr in 2000 to between roughly 1275 EJ/yr and 1500 EJ/yr in 2100. Dependence on conventional oil resources gradually decreases. However, a range of alternative fossil-based resources, such as synthetic fuels from coal and unconventional oil resources (e.g., tar sands and oil shales) are available and become economically viable. Fossil fuels provided almost 90% of global primary energy consumption in the year 2000, and they remain the dominant energy source in the three reference scenarios throughout the twenty-first century, supplying 70% to 80% of primary energy in 2100. Non-fossil fuel energy use grows over the century in all three reference scenarios. The range of contributions in 2100 is from 250 EJ/yr to 450 EJ/yr – an amount equaling roughly one-half to a little over global primary energy consumption today. *[Notes. i. Oil consumption includes that derived from tar sands and oil shales, and coal consumption includes that used to produce synthetic liquid and gaseous fuels. ii. Primary energy consumption from nuclear power and non-biomass renewable electricity are accounted for at the average efficiency of fossil-fired electric facilities, which vary over time and across scenarios. This long-standing convention means that, all other things being equal, increasing efficiency of fossil-electric energy lowers the contribution to primary energy from these sources.]*

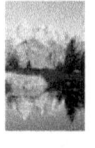

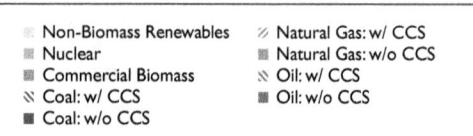

Legend:
- Non-Biomass Renewables
- Nuclear
- Commercial Biomass
- Coal: w/ CCS
- Coal: w/o CCS
- Natural Gas: w/ CCS
- Natural Gas: w/o CCS
- Oil: w/ CCS
- Oil: w/o CCS

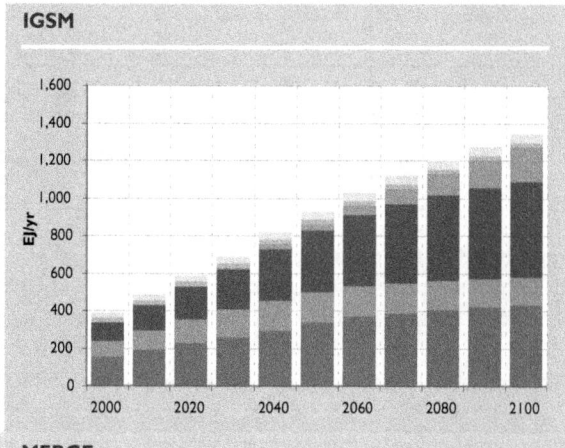

IGSM

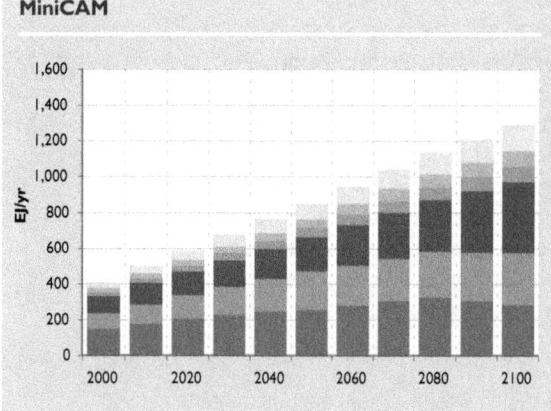

MiniCAM

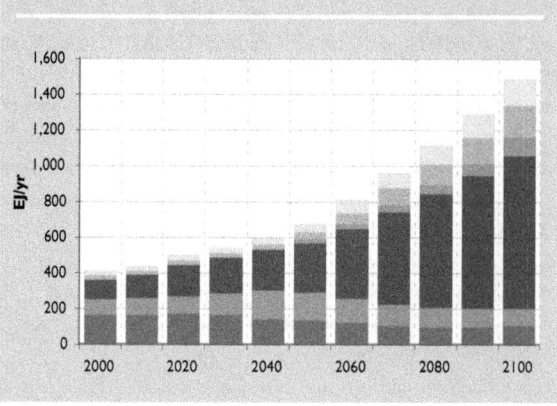

MERGE

The three reference scenarios show the implications of this increasing demand and the improved access to energy, with the ranges reflecting the variation among the scenarios from the three modeling groups. Global primary energy consumption rises substantially in all three reference scenarios, from about 400 EJ/yr in 2000 to between roughly 1275 EJ/yr and 1500 EJ/yr in 2100 (Figure TS.1). U.S. primary energy consumption also grows substantially, about 1¼ to 2½ times present levels by 2100 (Figure TS.2). Primary energy growth occurs despite continued improvements in the efficiency of energy use and energy production technologies. For example, the U.S. energy intensity – the ratio of primary energy consumption to economic output – declines 60% to 75%

between 2000 and 2100 across the three reference scenarios.

All three reference scenarios include an eventual reduction in the consumption of conventional oil resources. However, in all three, a range of alternative fossil-based resources, such as synthetic fuels from coal and unconventional oil resources (e.g., tar sands and oil shales), are available and become economically viable. Fossil fuels provided almost 90% of global primary energy in the year 2000, and they remain the dominant energy source in the three reference scenarios throughout the twenty-first century, supplying 70% to 80% of total primary energy in 2100.

Figure TS.2. U.S. Primary Energy Consumption Across Reference Scenarios (EJ/yr). U.S. primary energy consumption rises in all three reference scenarios, to roughly 1¼ to 2½ times present levels by 2100. This growth occurs despite continued improvements in the efficiency of energy use and production. U.S. energy intensity declines 60% to 75% between 2000 and 2100 in the reference scenarios. *[Notes. i. Oil consumption includes that derived from tar sands and oil shales, and coal consumption includes that used to produce synthetic liquid and gaseous fuels. ii. Primary energy consumption from nuclear power and non-biomass renewable electricity are accounted for at the average efficiency of fossil-fired electric facilities, which vary over time and across scenarios. This long-standing convention means that, all other things being equal, increasing efficiency of fossil-electric energy lowers the contribution to primary energy from these sources.]*

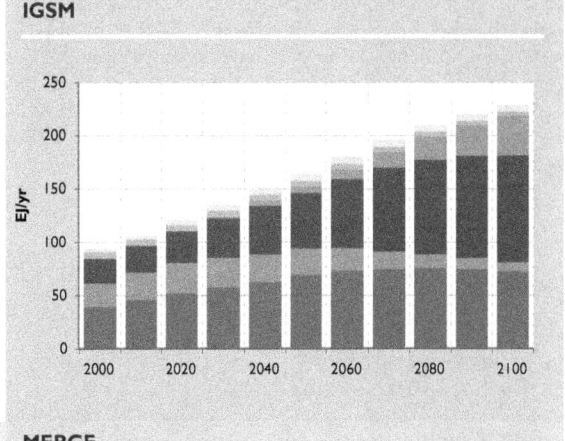

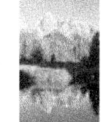

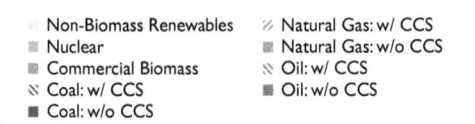

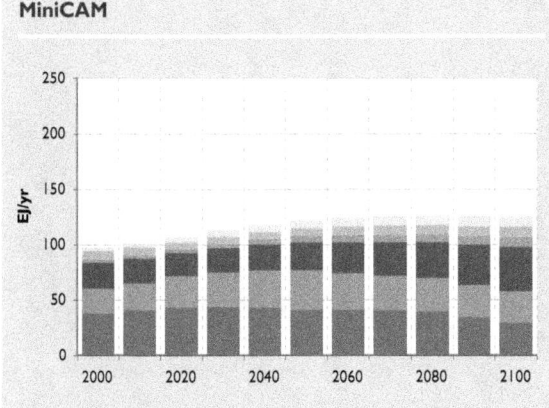

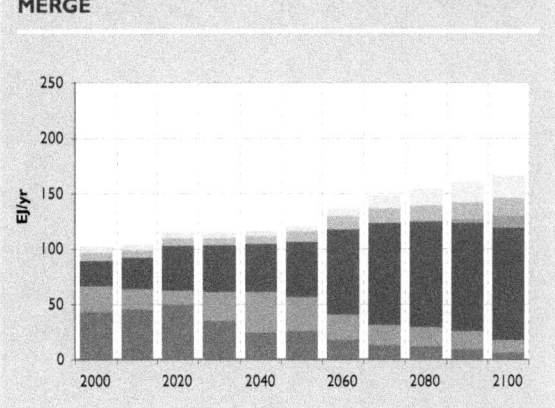

However, non-fossil fuel energy use also grows over the century in all three reference scenarios. Contributions to primary energy consumption in 2100 range from 250 EJ to 450 EJ – an amount equaling roughly ½ times to a little over global primary energy consumption today. Despite this growth, these sources never supplant fossil fuels, although they provide an increasing share of the total, particularly in the second half of the century.

Consistent with the characteristics of primary energy consumption, global and U.S. electricity production continues to rely on coal, although the contribution of coal varies among the reference scenarios (Figure TS.3 and Figure TS.4). The contribution of renewable and nuclear energy varies considerably in the different reference scenarios, depending on resource availability, technology, and non-climate policy considerations. For example, global nuclear

generation in the reference scenarios ranges from about 1½ times current levels (if non-climate concerns such as safety, waste, and proliferation constrain its growth as is the case in one reference scenario), to an expansion of almost an order of magnitude assuming relative economics as the only constraint.

In the reference scenarios, oil and natural gas prices rise through the century relative to year 2000 levels, whereas coal and electricity prices remain relatively stable. It should be emphasized, however, that the models used in this research were not designed to simulate short-term, fuel-price spikes, such as those that occurred in the 1970s, early 1980s, and more recently in 2005. Thus, price trends in the scenarios should be interpreted as multi-year averages.

As a combined result of all these influences, CO_2 emissions from fossil fuel combustion and

Figure TS.3. Global Electricity Production Across Reference Scenarios (EJ/yr). Global electricity production grows to over four times production levels in 2000 in all the reference scenarios. Global electricity production shows continued reliance on coal, although this contribution varies among the reference scenarios. The contribution of renewable energy and nuclear power varies considerably among the reference scenarios, depending on assumptions about resource availability, technology, and non-climate policy considerations. For example, global production of electricity from nuclear power in the reference scenarios ranges from about 1½ times current levels (if non-climate concerns such as safety, waste, and proliferation constrain its growth as is the case in one reference scenario), to an expansion of almost an order of magnitude assuming relative economics as the only constraint.

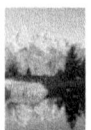

Legend:
- Non-Biomass Renewables
- Nuclear
- Commercial Biomass
- Coal: w/ CCS
- Coal: w/o CCS
- Natural Gas: w/ CCS
- Natural Gas: w/o CCS
- Oil: w/ CCS
- Oil: w/o CCS

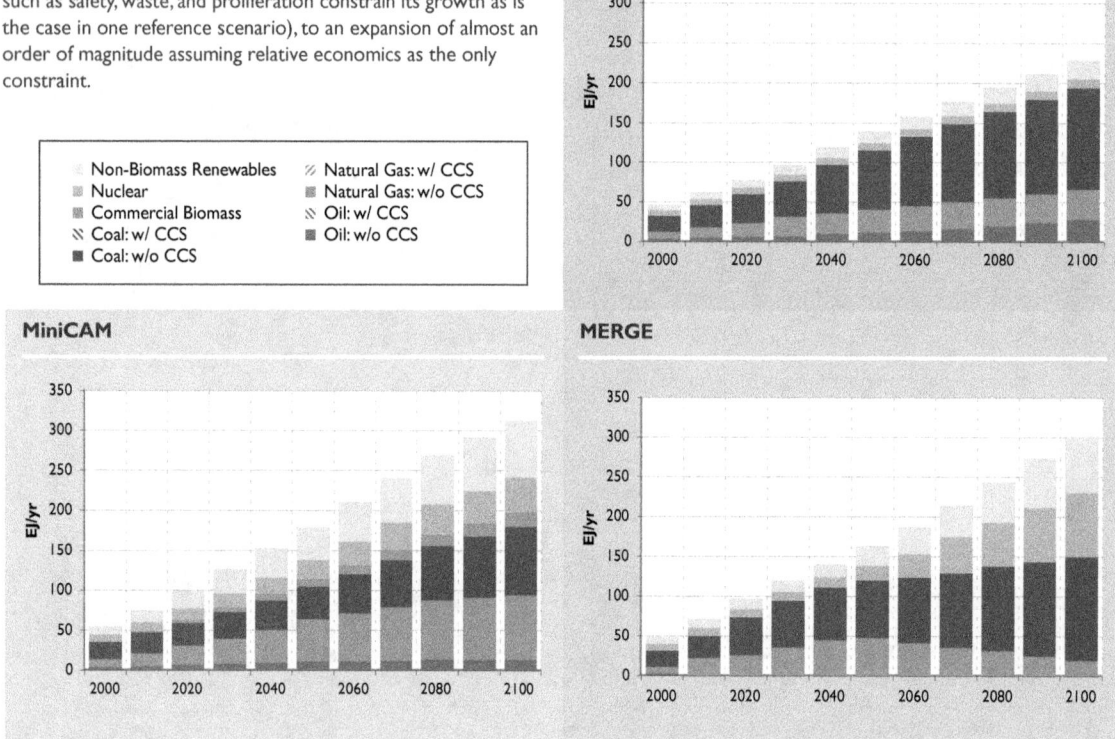

industrial processes in the reference scenarios increase from approximately 7 gigatonne carbon per year (GtC/yr) in 2000 to between 22.5 GtC/yr and 24.0 GtC/yr in 2100; that is, roughly 3 to 3½ times current levels (Figure TS.5). (Note that one tonne C is equivalent to 3.67 tonnes CO_2. See Box 3.2 for more on converting between units of carbon and units of CO_2.)

It is instructive to see how emissions are divided between industrialized countries (Annex 1) and developing countries (Non-Annex 1). Developing country emissions overtake those of developed countries in the 2020 to 2030 timeframe in the reference scenarios (Figure TS.6). This suggests the difficulty of stabilizing radiative forcing without developing-country participation. Indeed, even if developed countries were to reduce their emissions to zero, global involvement would still be necessary for stabilization.

The capacity of the ocean to absorb CO_2 differs among the three models. The ocean is a major sink for CO_2, and the rate at which the oceans take up CO_2 generally increases in the reference scenarios as concentrations rise early in the century. However, processes in the ocean can slow this rate of increase at high concentrations late in the century. Ocean uptake in the three reference scenarios is roughly 2 GtC/yr in 2000, rising to about 5 GtC/yr to 11 GtC/yr by 2100. The three ocean models behave more similarly in the stabilization scenarios; for example, the difference in ocean uptake between models, is less than 1 GtC/yr in 2100 under the most stringent stabilization level.

Two of the three participating models include sub-models of the exchange of CO_2 with the terrestrial biosphere, including the net uptake by plants and soils and the emissions from deforestation. In the reference scenarios from these

Figure TS.4. U.S. Electricity Production Across Reference Scenarios (EJ/yr). Continued dependence on coal for electricity generation is a feature of all three reference scenarios, with the degree of dependence varying among scenarios. Differences in the use of nuclear power reflect differing assumptions about the degree to which issues of safety, waste, and proliferation constrain its growth.

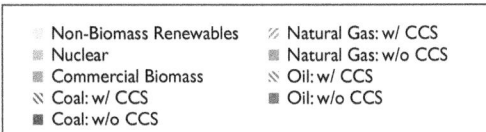

Non-Biomass Renewables · Natural Gas: w/ CCS
Nuclear · Natural Gas: w/o CCS
Commercial Biomass · Oil: w/ CCS
Coal: w/ CCS · Oil: w/o CCS
Coal: w/o CCS

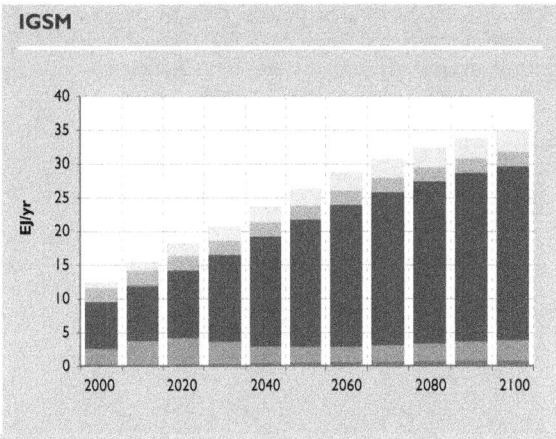

IGSM

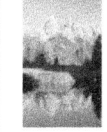

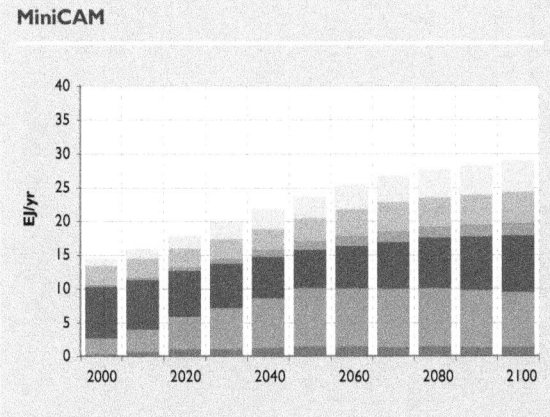

MiniCAM

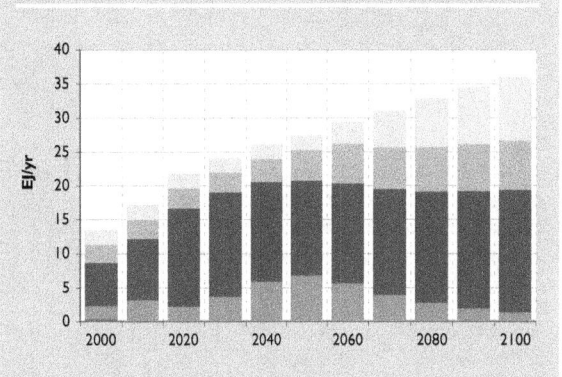

MERGE

modeling groups, the terrestrial biosphere acts as a small annual net sink (less than 1 GtC/yr) in 2000, increasing to an annual net sink of roughly 2 GtC/yr to 3 GtC/yr by the end of the century. The third modeling group assumed a zero net exchange. Changes in emissions from terrestrial systems over time in the reference scenarios reflect assumptions about human activity (including a decline in deforestation) as well as increased CO_2 uptake by vegetation as a result of the positive effect of CO_2 on plant growth. There remains substantial uncertainty about this carbon fertilization effect and its evolution under a changing climate.

Although this Technical Summary focuses on the most important anthropogenic GHG, CO_2, the scenarios considered a number of other GHGs (CH_4, N_2O SF_6, PFCs, and HFCs), which are emitted from various sources, including agriculture, waste management, biomass burning, fossil fuel production and consumption, and a number of industrial activities. Future global anthropogenic emissions of CH_4 and N_2O vary widely among the reference scenarios, ranging from flat or declining emis-

sions to increases of 2 to 2½ times present levels. These differences reflect differing assumptions about technological opportunities and about whether current emissions rates will be reduced significantly for non-climate reasons, such as air pollution control and/or higher natural gas prices that would further stimulate the capture of CH_4 emissions for its fuel value.

Increases in emissions from the global energy system and other human activities lead to higher atmospheric GHG concentrations and radiative forcing. These increases are moderated by natural biogeochemical removal processes. As a result, GHG concentrations rise substantially over the century in the reference scenarios. By 2100, CO_2 concentrations range from about 700 ppmv to 900 ppmv, up from 365 ppmv in 1998. CH_4 concentrations in 2100 range from 2000 ppbv to 4000 ppbv, up from 1745 ppbv in 1998, and N_2O concentrations in 2100 range from about 375 ppbv to 500 ppbv, up from 314 ppbv in 1998.

As a result, radiative forcing in 2100 ranges from 6.4 W/m² to 8.6 W/m² from preindustrial, up from a little over 2 W/m² today. The non-CO_2

Figure TS.5. Global Emissions of CO₂ from Fossil Fuels and Industrial Sources [CO₂ from land-use change excluded] Across Reference Scenarios (GtC/yr). Global emissions of CO₂ from fossil fuel combustion and other industrial sources, mainly cement production, increase over the century in all three reference scenarios. By 2100 emissions reach 22.5 GtC/yr to 24.0 GtC/yr.

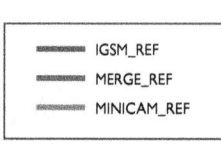

IGSM_REF
MERGE_REF
MINICAM_REF

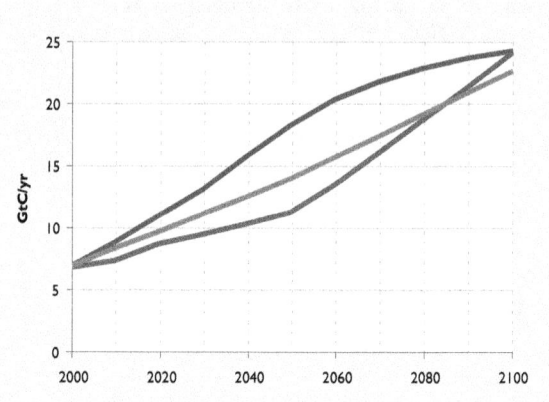

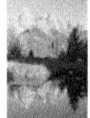

Figure TS.6. Global Emissions of Fossil Fuel and Industrial CO₂ by Annex I and Non-Annex I Countries Across Reference Scenarios (GtC/yr). Emissions of fossil fuel and industrial CO₂ from the Non-Annex I countries exceed those from the Annex I countries in all three reference scenarios by 2030 or earlier. Non-Annex I emissions continue to grow rapidly in two of the reference scenarios, such that their emissions are on the order of twice the level of Annex I by 2100. Emissions do not continue to diverge in the third reference scenario, due in part to relatively slower economic growth in Non-Annex I regions, faster growth in Annex I, and increased emissions in Annex I as they become producers and exporters of shale oil, tar sands, and synthetic fuels from coal.

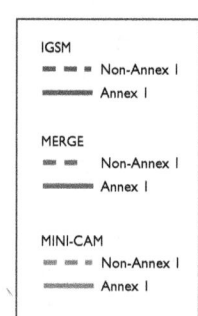

IGSM
- - - Non-Annex I
——— Annex I

MERGE
- - - Non-Annex I
——— Annex I

MINI-CAM
- - - Non-Annex I
——— Annex I

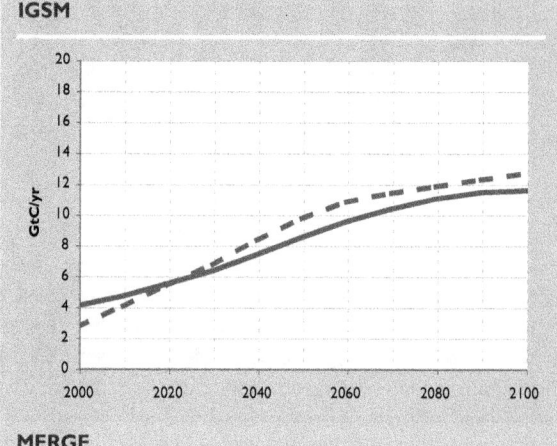

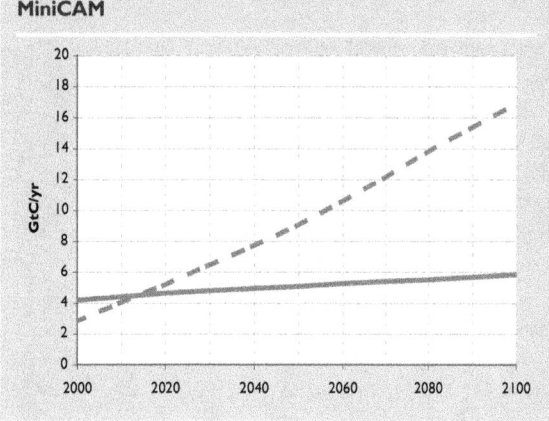

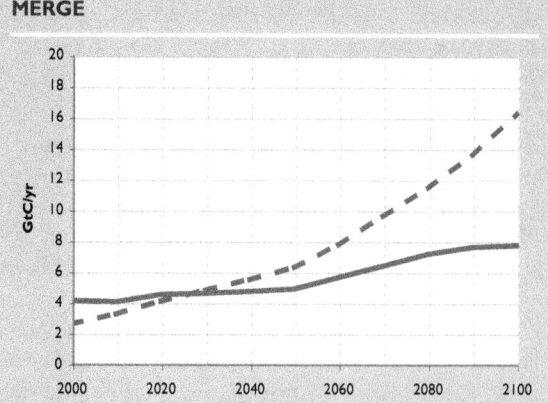

Figure TS.7. Radiative Forcing by Gas Across Reference Scenarios (W/m² from preindustrial).
CO_2 accounts for 75% to 80% of the radiative forcing in 2100 in the three reference scenarios. Total radiative forcing in 2100 from all the GHGs considered in this research ranges from about 6.4 W/m² to 8.6 W/m².

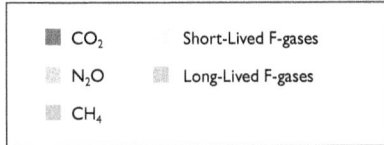

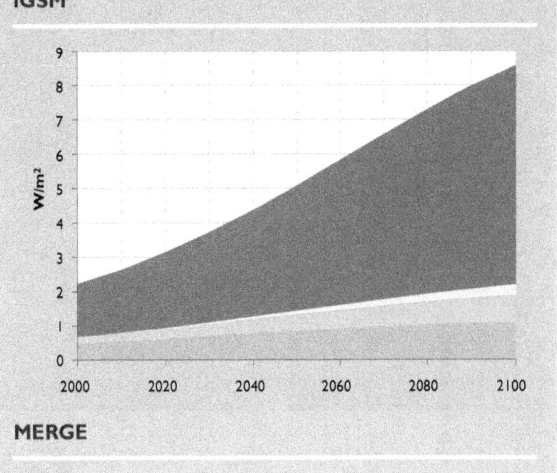

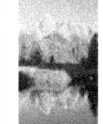

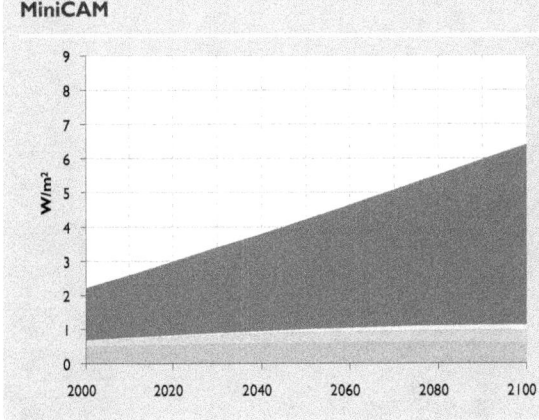

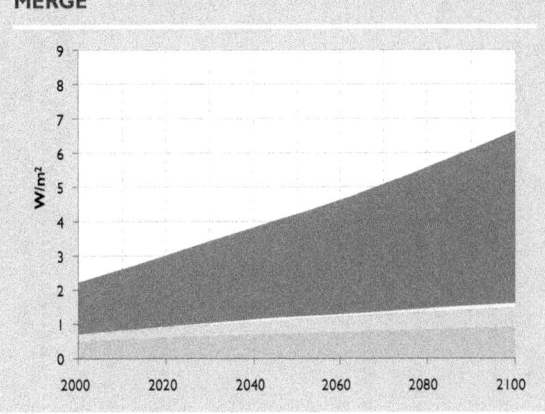

GHGs account for about 20% to 25% of radiative forcing by the end of the century (Figure TS.7).

Stabilization Scenarios

Important assumptions underlying the stabilization scenarios include the flexibility that exists in a policy design, as represented by the modeling groups, to seek out least cost options for emissions control regardless of where they occur, what substances are controlled, or when they occur. This set of conditions is referred to as *where*, *what*, and *when* flexibility. Equal marginal costs of abatement among regions, across time (taking into account discount rates and the lifetimes of substances), and among substances (taking into account their relative warming potential and different lifetimes) will, under specified conditions, lead to least-cost abatement. Each modeling group applied an economic instrument that priced GHGs in a manner consistent with the group's interpretation of *where*, *what* and *when* flexibility. The economic characteristics of the scenarios therefore assume a policy designed with the intent of achieving the

required reductions in GHG emissions in a least-cost way. Key implications of these assumptions are that: (1) all nations proceed together in restricting GHG emissions from 2012 and continue together throughout the century, and that the same marginal cost is applied across sectors (*where* flexibility); (2) the marginal cost of abatement rises over time based on each modeling group's interpretation of *when* flexibility, with the effect of linking emissions mitigation efforts over the time horizon of the scenarios; and (3) stabilization of radiative forcing is achieved by combining control of all GHGs, with differences in how modeling groups compared them (*what* flexibility).

Although these assumptions are convenient for analytical purposes, to gain an impression of the implications of stabilization, they are idealized versions of possible outcomes. For the abatement costs in these scenarios to be representative of actual abatement costs would require, among other things, that a negotiated international agreement include these flexibility mechanisms. Failure in that regard could have a

Figure TS.8. Global Primary Energy Consumption by Fuel Across Scenarios (EJ/yr). The global energy system undergoes a significant transformation in the stabilization scenarios from all three modeling groups. This transformation begins earlier the more stringent the radiative forcing stabilization level, and would continue into the next century for all stabilization levels. The transformation includes: reduction in energy use, increased use of carbon-free sources of energy such as biomass, other renewables, and nuclear power; and the addition of CCS. The contribution of each of these varies among the models reflecting different assumptions about cost and performance, policy, and resource limits. *[Notes. i. Oil consumption includes that derived from tar sands and oil shales, and coal consumption includes that used to produce synthetic liquid and gaseous fuels. ii. Primary energy consumption from nuclear power and non-biomass renewable electricity are accounted for at the average efficiency of fossil-fired electric facilities, which vary over time and across scenarios. This long-standing convention means that, all other things being equal, increasing efficiency of fossil-electric energy lowers the contribution to primary energy from these sources.]*

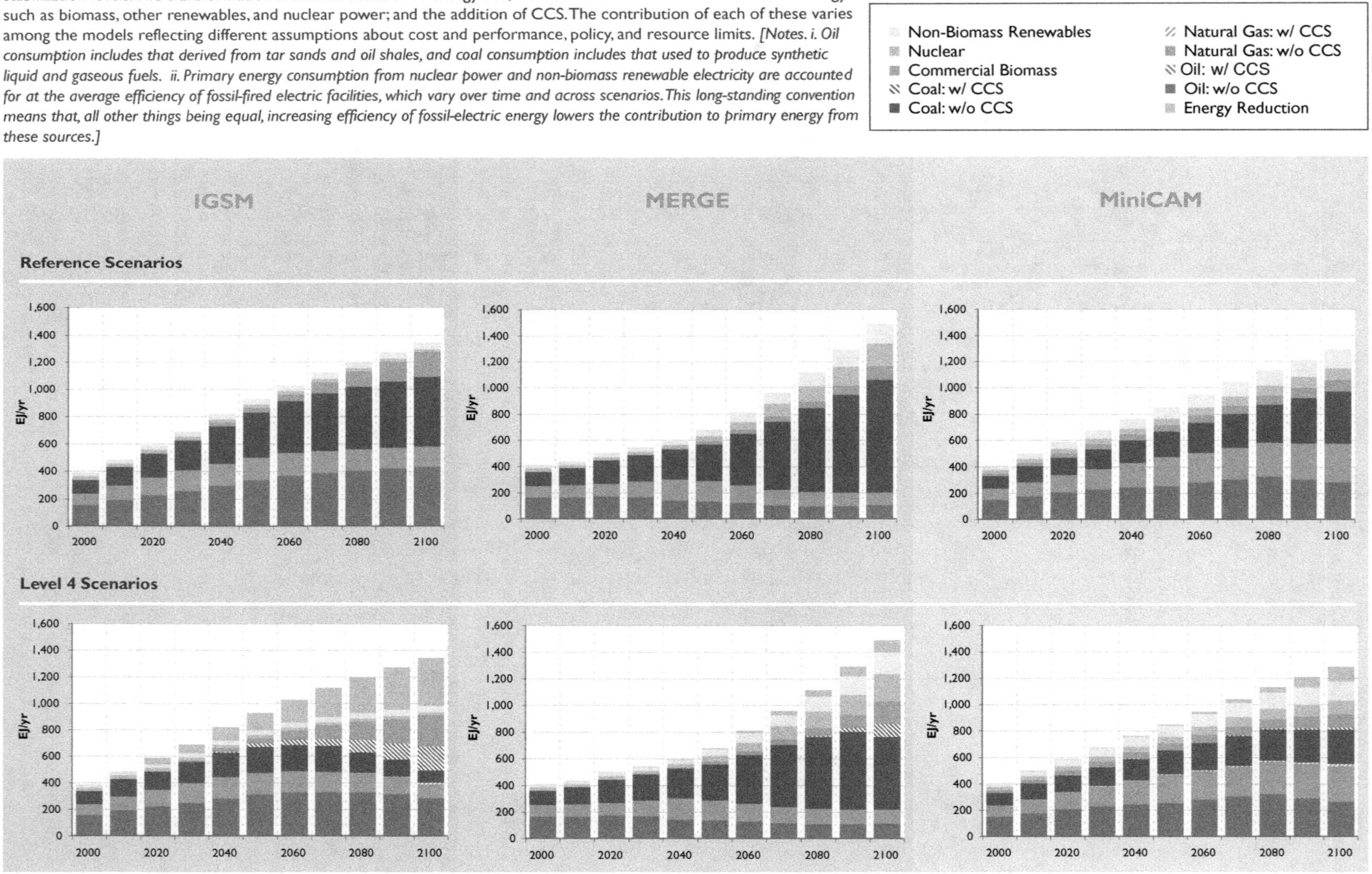

Legend:
- Non-Biomass Renewables
- Nuclear
- Commercial Biomass
- Coal: w/ CCS
- Coal: w/o CCS
- Natural Gas: w/ CCS
- Natural Gas: w/o CCS
- Oil: w/ CCS
- Oil: w/o CCS
- Energy Reduction

IGSM   MERGE   MiniCAM

Reference Scenarios

Level 4 Scenarios

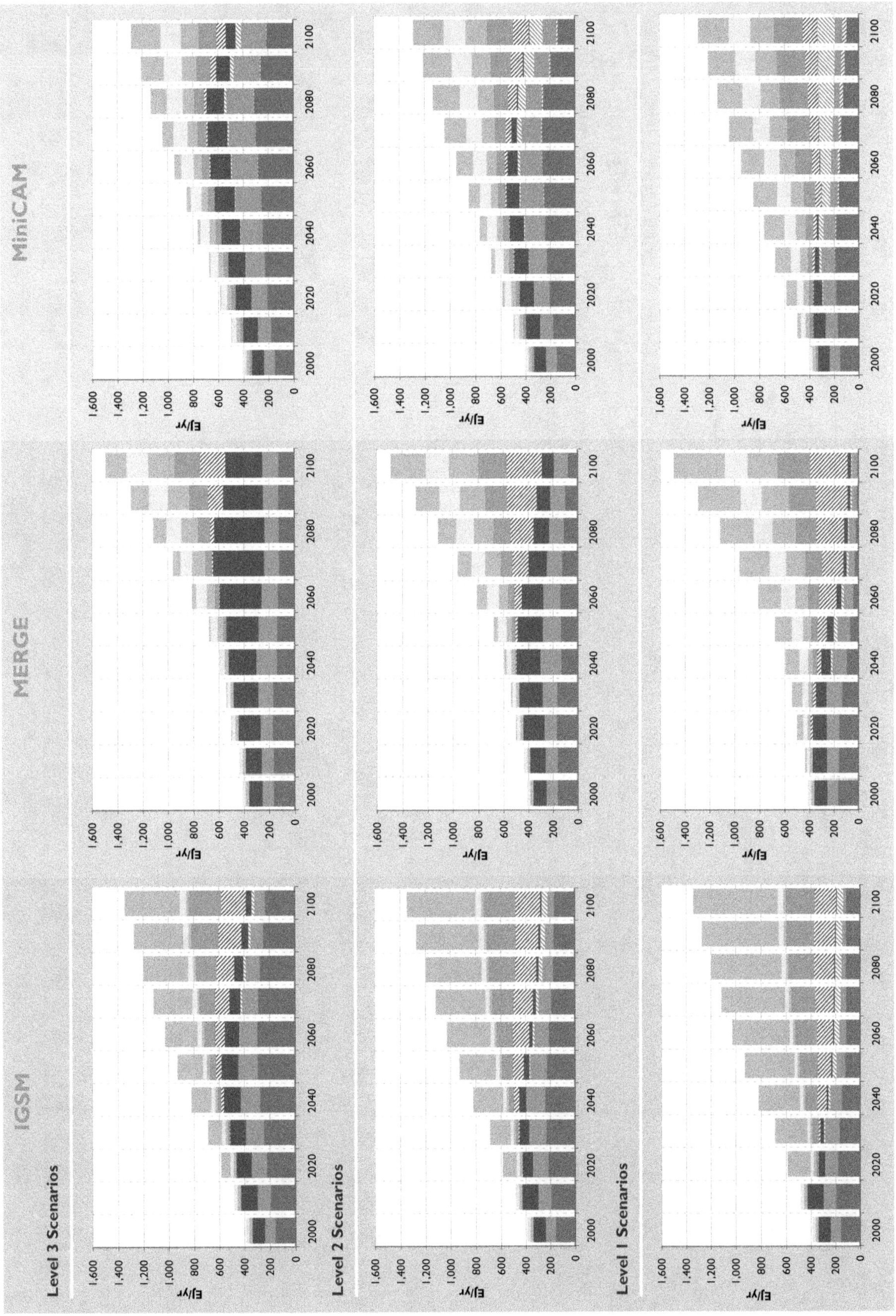

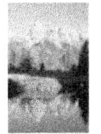

Figure TS.9. U.S. Primary Energy Consumption by Fuel Across Scenarios (EJ/yr). The U.S. energy system undergoes a significant transformation in the stabilization scenarios similar to the transformation in the global energy system. One difference, not obvious in this figure, is the transformation from conventional oil and gas to synthetic fuel production derived from shale oil or coal. One model (IGSM) includes heavy use of shale oil in the reference with some coal gasification, whereas another (MERGE) includes primarily synthetic liquid and gaseous fuels derived from coal. The third (MiniCAM) includes an intermediate mix of both. [Notes. *i.* Oil consumption includes that derived from tar sands and oil shales, and coal consumption includes that used to produce synthetic liquid and gaseous fuels. *ii.* Primary energy consumption from nuclear power and non-biomass renewable electricity are accounted for at the average efficiency of fossil-fired electric facilities, which vary over time and across scenarios. This long-standing convention means that, all other things being equal, increasing efficiency of fossil-electric energy lowers the contribution to primary energy from these sources.]

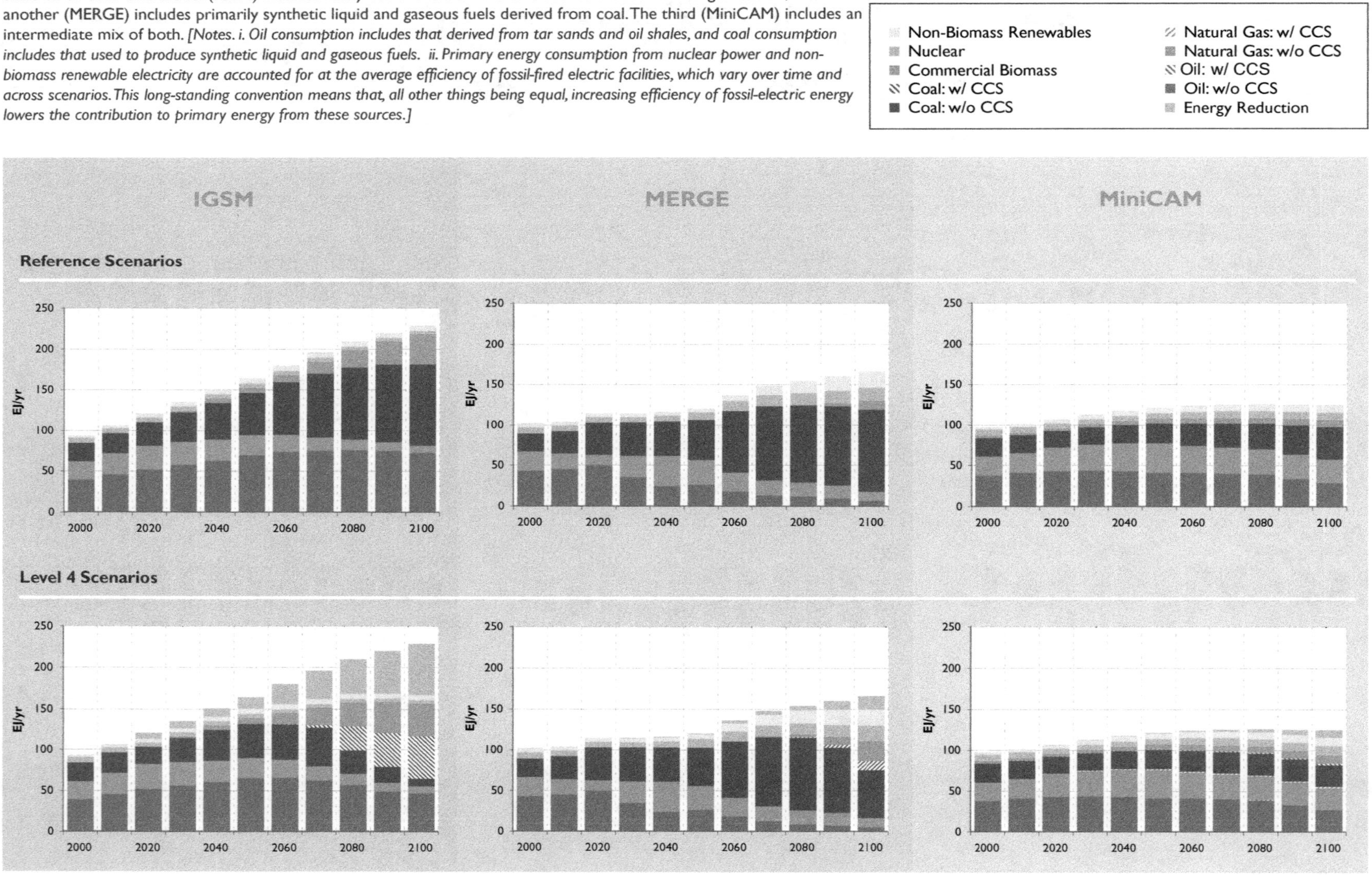

Reference Scenarios

Level 4 Scenarios

IGSM

MERGE

MiniCAM

Non-Biomass Renewables
Nuclear
Commercial Biomass
Coal: w/ CCS
Coal: w/o CCS

Natural Gas: w/ CCS
Natural Gas: w/o CCS
Oil: w/ CCS
Oil: w/o CCS
Energy Reduction

EJ/yr

2000 2020 2040 2060 2080 2100

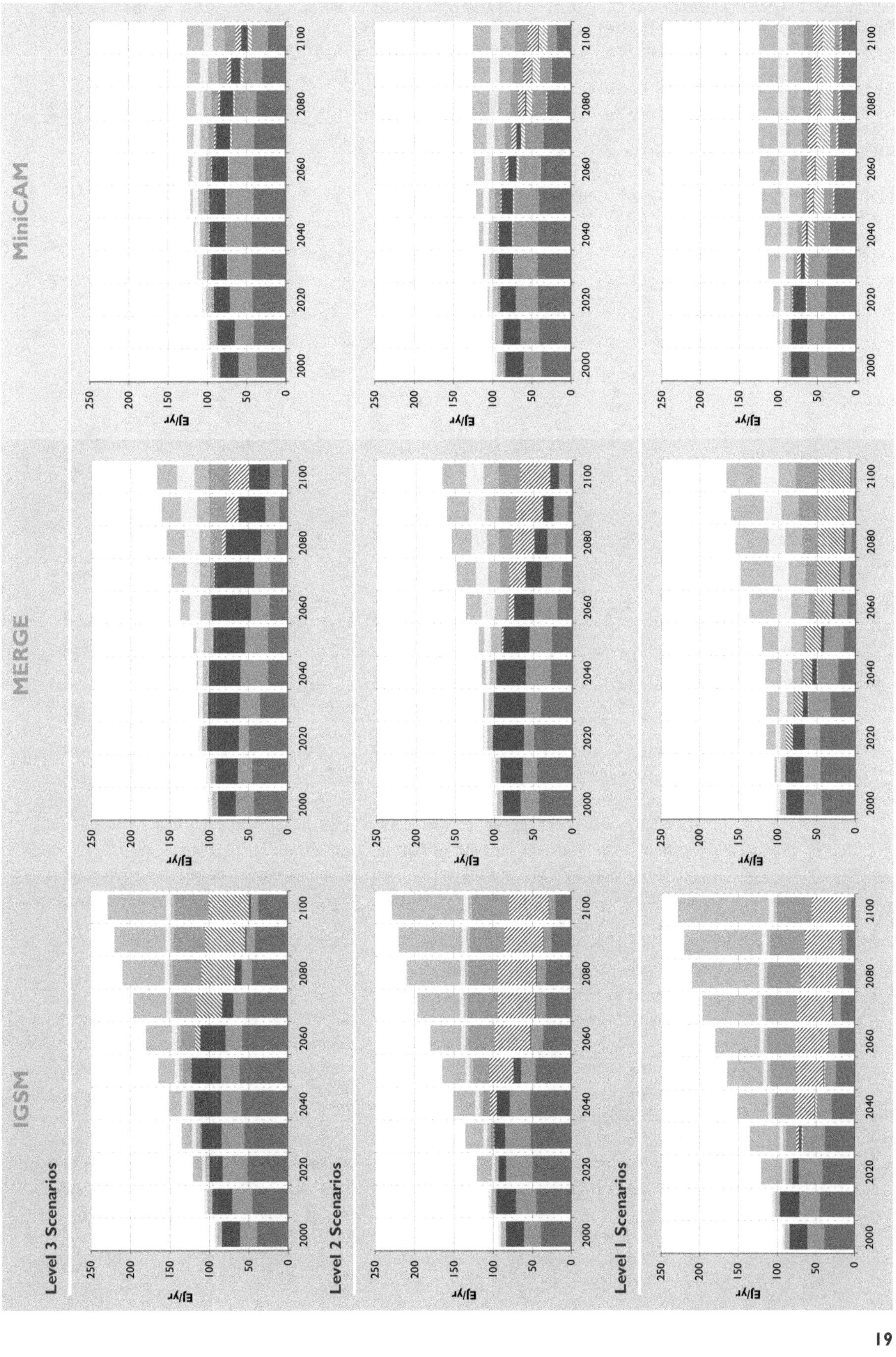

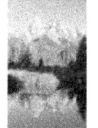

IGSM
MERGE
MiniCAM
Level 3 Scenarios
Level 2 Scenarios
Level 1 Scenarios
EJ/yr
250
200
150
100
50
0
2000
2020
2040
2060
2080
2100

Figure TS.10 Global Emissions of CO$_2$ from Fossil and other Industrial Sources Across Scenarios (GtC/yr).
The tighter the constraint on radiative forcing, the faster carbon emissions must decline from those in the reference scenarios. This is

because the stabilization level defines a long-term carbon budget; that is, the remaining amount of carbon that can be emitted in the future. The gradual deflection of the emissions from the reference reflects the assumption of *when* flexibility, with carbon prices rising gradually. Under the most stringent radiative forcing stabilization levels, CO$_2$ emissions begin to decline immediately or within a matter of decades. Under less stringent radiative forcing stabilization levels, CO$_2$ emissions do not peak until late in the century or beyond, and they are 1½ to over 2½ times today's levels in 2100.

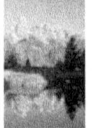

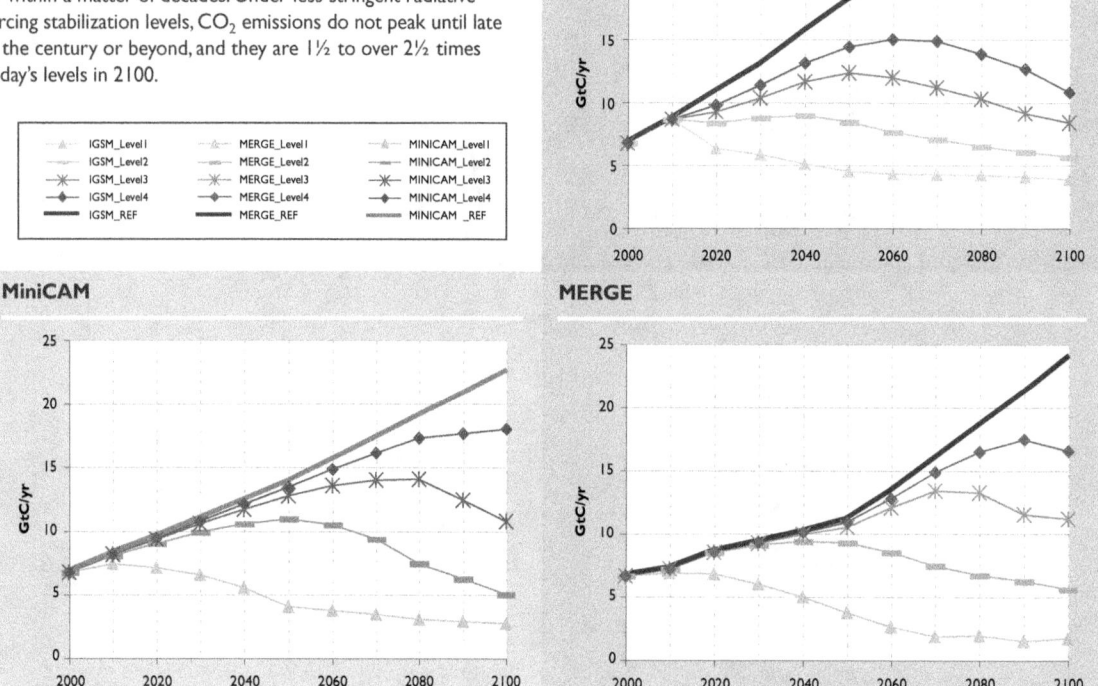

substantial effect on the difficulty of achieving any of the stabilization levels considered in this research. For example, a delay in the participation of some large countries would require greater effort by the others, and policies that impose differential burdens on different sectors without mechanisms to allow for equalizing marginal costs across sectors can result in a many-fold increase in the cost of any environmental gain. Therefore, *it is important to view these result as scenarios under specified conditions, not as predictions or best-judgment forecasts of the most likely outcome within the national and international political system.* Further, none of the scenarios considered the extent to which variation from these least-cost rules might be improved upon given interactions with existing taxes, technology spillovers, or other non-market externalities.

If the developments in the three reference scenarios were to occur, concerted efforts to reduce

GHG emissions would be required to stabilize radiative forcing at the levels considered in this research. Such limits would shape technology deployment throughout the century and have important economic consequences. The stabilization scenarios demonstrate that there is no single technology pathway consistent with a given level of radiative forcing. Furthermore, there are other possible pathways than those considered in this research.

Stabilization of radiative forcing at the levels examined in this research would require a substantially different energy system globally, and in the U.S., than what emerges in the reference scenarios. The degree and timing of change in the global energy system depends on the level at which radiative forcing is stabilized (Figure TS.8 and Figure TS.9). The lower the radiative forcing stabilization level, the larger the scale of change in the global energy system relative to the reference scenario required over the coming century and the sooner those changes would need to occur.

Across the stabilization scenarios, the energy system relies more heavily on non-fossil energy sources, such as nuclear, solar, wind, biomass, and other renewable energy forms, than in the associated reference scenarios. The stabilization scenarios differ in the degree to which these technologies are deployed, depending on assumptions about: technological improvements; the ability to overcome obstacles, such as intermittency in the case of solar and wind power, or safety, waste, and proliferation issues in the case of nuclear power; and the policy environment surrounding these technologies. Energy consumption, while still higher than today's levels, is lower in the stabilization scenarios than in the reference scenarios.

CCS is widely deployed in the stabilization scenarios because each modeling group assumed that the technology can be successfully developed and that concerns about storing large amounts of carbon do not impede its expansion. Removal of this assumption would make the stabilization levels more difficult to achieve and would lead to greater demand for low-carbon sources such as renewable energy and nuclear power, to the extent that growth of these other sources is not otherwise constrained.

Significant fossil fuel use continues across the stabilization scenarios, because stabilization allows for some level of carbon emissions through 2100, and because of the presence of CCS technology in all the stabilization scenarios.

Increased use is made of biomass energy crops in all the stabilization scenarios, the contribution of which is ultimately limited by competition with agriculture and forestry. One modeling group examined the importance of valuing terrestrial carbon similarly to the way fossil fuel carbon is valued in stabilization scenarios. It was found that important interactions between large-scale deployment of commercial bioenergy crops and land use occurred to the detriment of unmanaged ecosystems when no economic value was placed on carbon in terrestrial systems.

Across the stabilization scenarios, the scale of the emissions reductions required relative to the reference scenario increases over time, with the bulk of emissions reductions taking place in the second half of the century. But emissions reductions occur in the first half of the century in every stabilization scenario (Figure TS.10).

The 2100 time horizon of this research limited examination of the ultimate stabilization requirements. Further reductions in CO_2 emissions after 2100 would be required in all of the stabilization scenarios, because stabilization of radiative forcing at any of the levels considered in this research requires human emissions of CO_2 in the long term to be essentially halted. Despite the fact that much of the carbon emissions will eventually make its way into oceans and terrestrial sinks, some will remain in the atmosphere for thousands of years. Only CCS can allow continued burning of fossil fuels. Higher radiative forcing limits can delay the point in time at which emissions must be reduced toward zero, but this requirement must ultimately be met.

Fuel sources and electricity generation technologies change substantially, both globally and in the U.S., in the stabilization scenarios compared to the reference scenarios. There are a variety of technological options in the electricity sector that reduce carbon emissions in these scenarios (Figure TS.11 and Figure TS.12).

By the end of the century, electricity produced by conventional fossil technology that freely emits CO_2 is reduced in the stabilization scenarios relative to reference scenarios. Electricity production from technologies that emit CO_2 varies substantially with the stabilization level; in the most stringent stabilization scenarios, electricity production from these technologies is reduced toward zero.

The economic effects of stabilization are substantial in many of the stabilization scenarios, although much of this cost is borne later in the century. As noted earlier, each of the modeling groups assumed that a global policy was implemented after 2012, with universal participation by the world's nations, and that the time path of reductions approximated a least-cost solution.

Figure TS.11. Global Electricity Production by Fuel Across Scenarios (EJ/yr). Various electricity technology options could be competitive in the future, and different assumptions regarding their relative economic viability, reliability, and resource availability lead to considerably different scenarios of the global electricity sector in reference and stabilization scenarios across modeling groups. One reference scenario includes relatively little change in the electricity sector mix in the reference scenario. The other two reference scenarios include more substantial transformations from the present. In all scenarios, large changes from reference are required to stabilize radiative forcing at the levels considered in this research. In most cases, the relative proportion of electricity in energy consumption increases in the stabilization scenarios, so the relative reductions in electricity production are generally smaller than for primary energy.

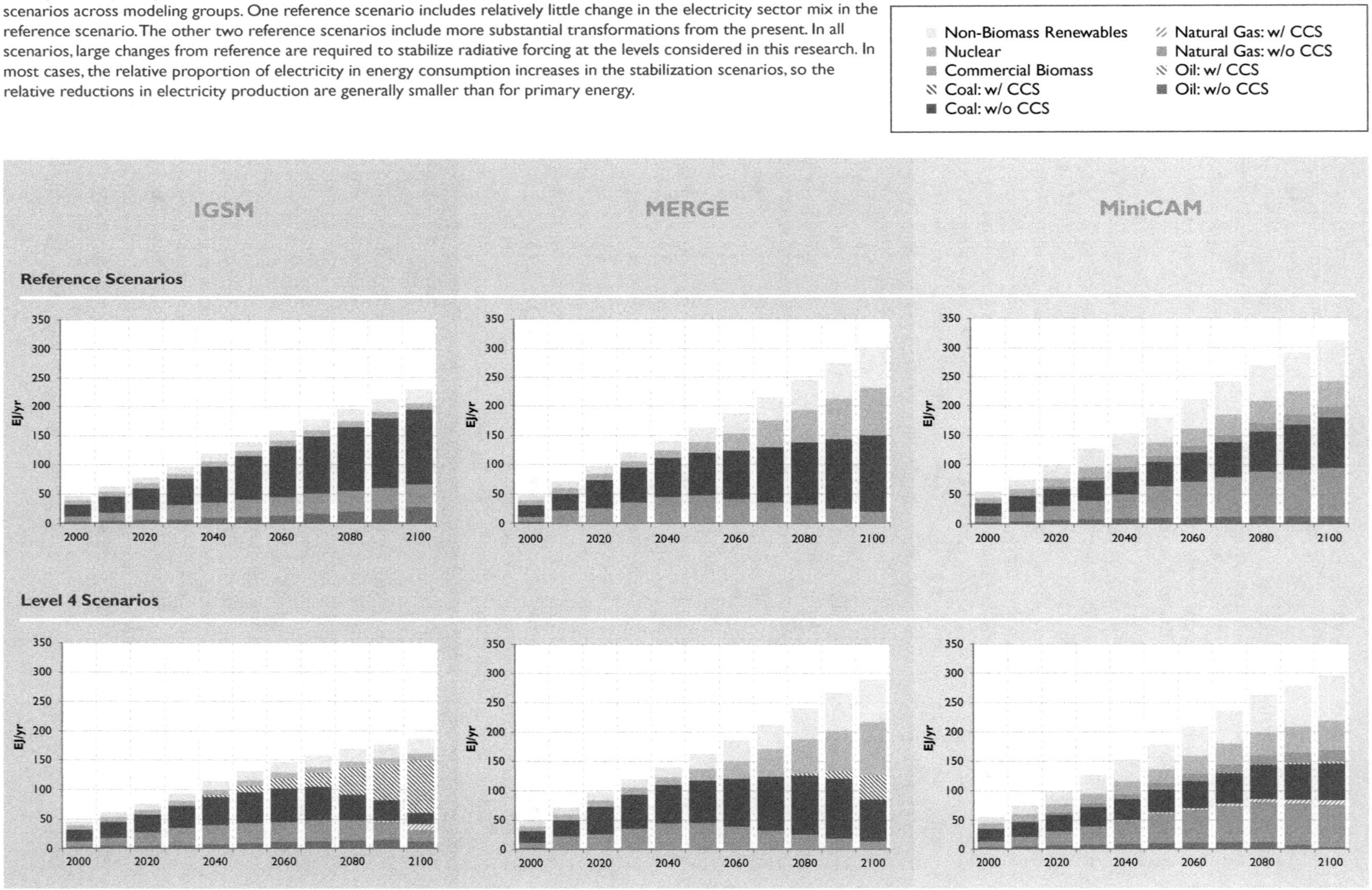

Non-Biomass Renewables
Nuclear
Commercial Biomass
Coal: w/ CCS
Coal: w/o CCS
Natural Gas: w/ CCS
Natural Gas: w/o CCS
Oil: w/ CCS
Oil: w/o CCS

IGSM

MERGE

MiniCAM

Reference Scenarios

Level 4 Scenarios

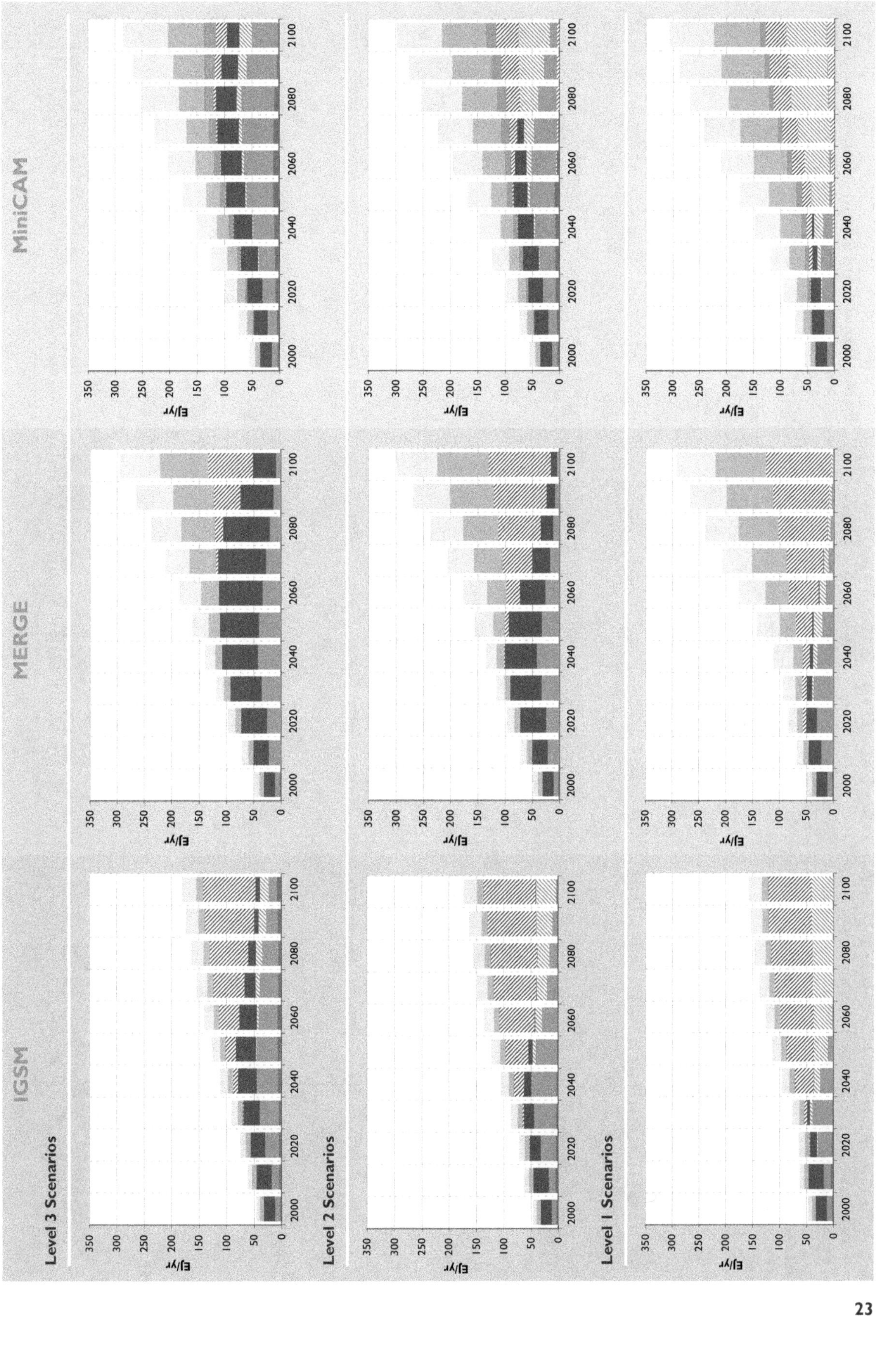

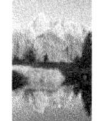

IGSM

MERGE

MiniCAM

Level 3 Scenarios

Level 2 Scenarios

Level 1 Scenarios

EJ/yr

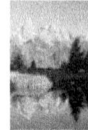

Figure TS.12. U.S. Electricity Production by Fuel Across Scenarios (EJ/yr). U.S. electricity generation sources and technologies are substantially transformed to meet the four radiative forcing stabilization levels. CCS figures in the stabilization scenarios from all three modeling groups, but the contribution of other sources and technologies and the total amount of electricity used differ substantially. In most cases, the relative proportion of electricity in energy consumption increases in the stabilization scenarios, so the relative reductions in electricity production are generally smaller than for primary energy. In one scenario (MiniCAM Level 1), electricity production in the U.S. increases under stabilization in the second half of the century.

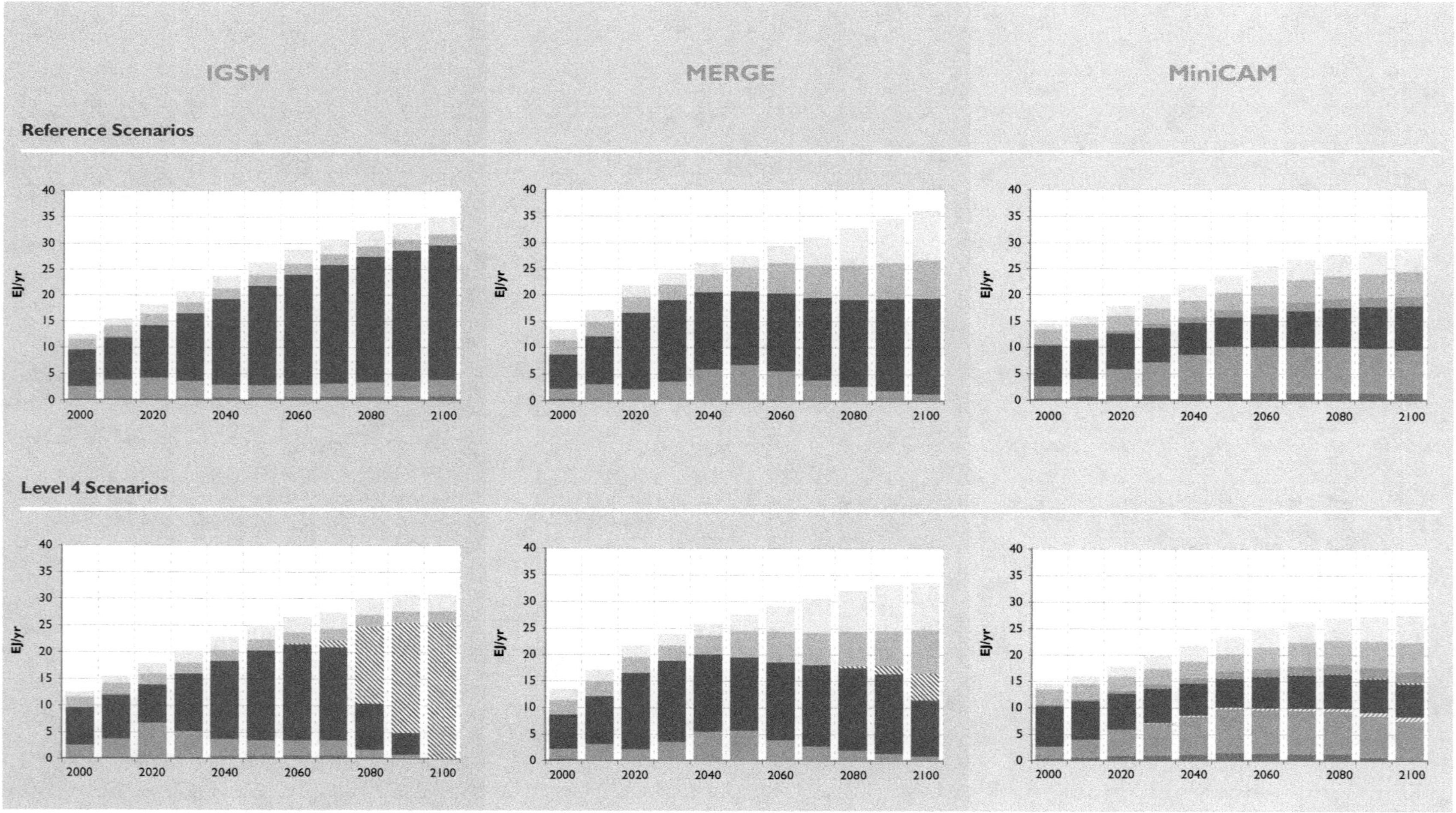

Legend:
- Non-Biomass Renewables
- Nuclear
- Commercial Biomass
- Coal: w/ CCS
- Coal: w/o CCS
- Natural Gas: w/ CCS
- Natural Gas: w/o CCS
- Oil: w/ CCS
- Oil: w/o CCS

IGSM

MERGE

MiniCAM

Reference Scenarios

Level 4 Scenarios

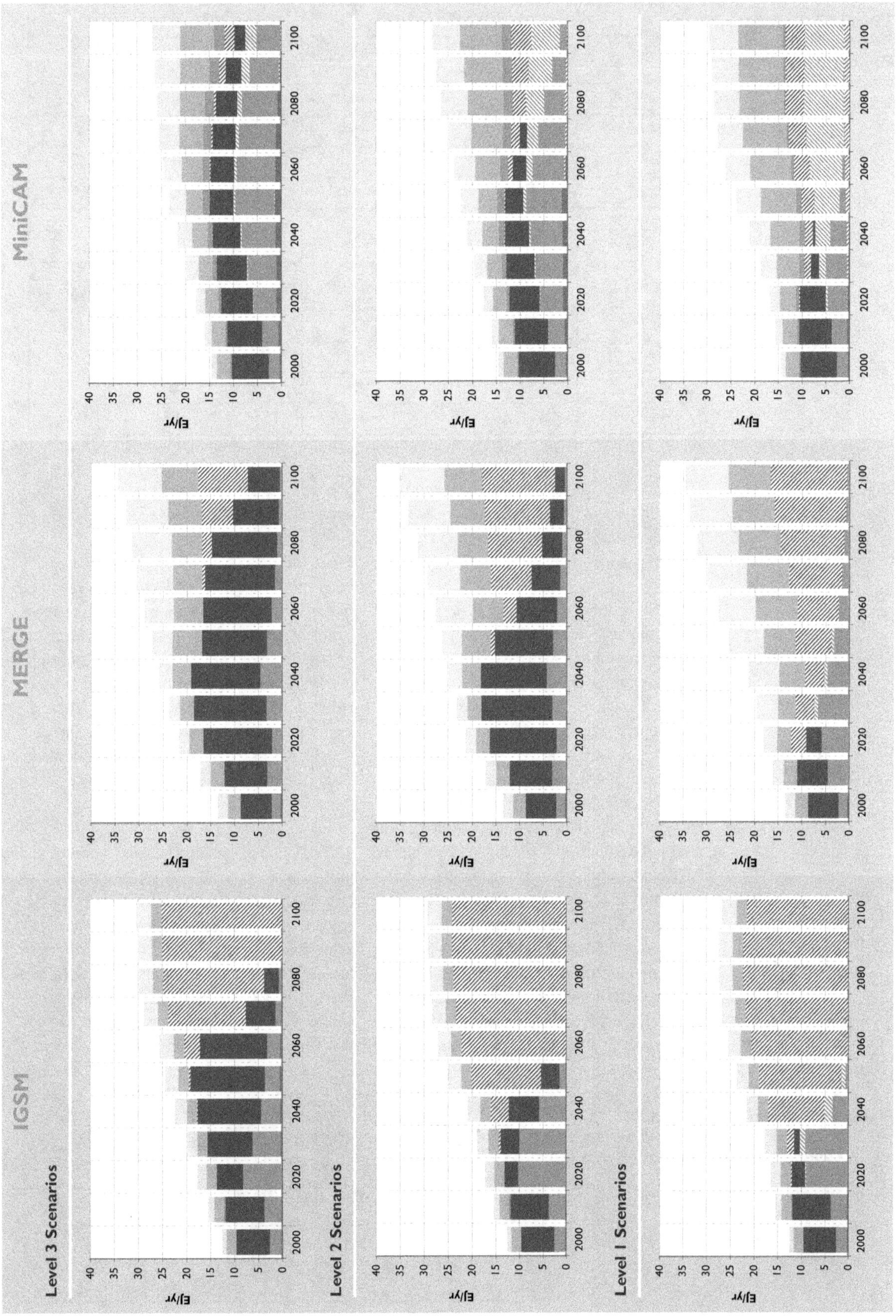

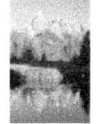

IGSM   **MERGE**   **MiniCAM**

Level 3 Scenarios

Level 2 Scenarios

Level 1 Scenarios

Table TS.3. Carbon Prices at Various Points in Time for the Stabilization Scenarios

Stabilization Level	2020 ($/tonne C)			2030 ($/tonne C)		
	IGSM	MERGE	MiniCAM	IGSM	MERGE	MiniCAM
Level 4	$18	$1	$1	$26	$2	$2
Level 3	$30	$2	$4	$44	$4	$7
Level 2	$75	$8	$15	$112	$13	$26
Level 1	$259	$110	$93	$384	$191	$170

Stabilization Level	2050 ($/tonne C)			2100 ($/tonne C)		
	IGSM	MERGE	MiniCAM	IGSM	MERGE	MiniCAM
Level 4	$58	$6	$5	$415	$67	$54
Level 3	$97	$11	$19	$686	$127	$221
Level 2	$245	$36	$69	$1,743	$466	$420
Level 1	$842	$574	$466	$6,053	$609	$635

These assumptions of where, when, and *what* flexibility lower the economic consequences of stabilization relative to what they might be with other implementation approaches.

The stabilization scenarios follow a pattern where, in most scenarios, the carbon price rises steadily over time (Table TS.3), providing an opportunity for the energy system to adjust gradually. Although the general shape of the carbon price trajectory over time is similar across the models, the carbon prices vary substantially across the models. For example, for the less stringent stabilization levels two of the modeling groups produced scenarios with carbon prices of $10 or below per tonne of carbon in 2020, with carbon prices rising to roughly $100 per tonne in 2020 at the most stringent stabilization level. The scenarios from the third modeling group show higher initial carbon prices in 2020, ranging from around $20 for the least stringent stabilization level to over $250 for the most stringent stabilization level. (Note that $100/tonne C is equivalent to $27/tonne CO_2.

See Box 3.2 for more on converting between units of carbon and units of CO_2.)

These differences in carbon prices, along with other model features, lead to similar variation in the costs of stabilization. Under the most stringent radiative forcing stabilization level, for example, gross world product (aggregating country figures using market exchange rates) is reduced in 2050 by around 1% in the scenarios from two of the modeling groups and approximately 5% in the scenario from the third. In 2100 it is reduced by less than 2% in two of the scenarios and over 16% in the third.

The variation in carbon prices and reductions in gross world product is attributable to many factors, but two are most prominent. First, the amount that CO_2 emissions must be reduced to achieve stabilization differs between the scenarios from the different modeling groups (Table TS.4), because of differing assumptions regarding economic growth and other factors that determine emissions in the reference sce-

Table TS.4. Cumulative Emissions Reductions from the Reference Scenarios Across Models in the Stabilization Scenarios (GtC through 2100)

	IGSM	MERGE	MiniCAM
Level 4	472	112	97
Level 3	674	258	267
Level 2	932	520	541
Level 1	1172	899	934

Figure TS.13. Relationship Between Carbon Price and Percentage Emissions Reductions in 2050 and 2100.
The relationship between carbon price and percentage reduction in emissions is similar among the models in 2050. In 2100, the relationship between carbon price and percentage reduction in emissions diverges across the models, due in large part to different assumptions regarding the technologies available to facilitate emissions reductions late in the century. [Note. CO_2 emissions vary across the reference scenarios from the three modeling groups, so that similar percentage reductions, as shown in this figure, imply differing levels of total emissions reduction.]

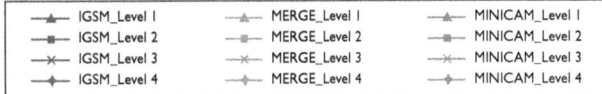

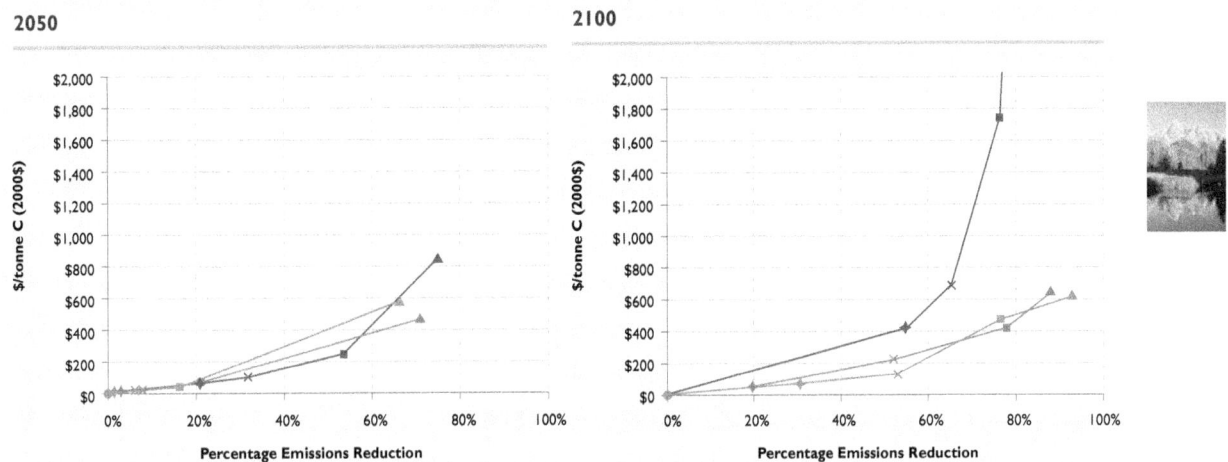

narios; levels of CO_2 uptake by the oceans and terrestrial biosphere; and availability of control for non-CO_2 GHGs.

Second, the modeling groups chose different assumptions regarding the technologies available for emissions reductions, particularly in the second half of the century. Most prominent are differences in assumptions about technologies to shift final energy demand to low-carbon sources such as biofuels and low-carbon electricity or hydrogen, in transportation, industrial and buildings end uses. The differences in technological assumptions among the modeling groups is reflected in the relationship between carbon prices and percentage abatement (Figure TS.13), a form of marginal abatement cost curve, for the three models in 2050 and 2100. The scenarios from the three modeling groups exhibit very similar behavior through 2050, but different assumptions about technological options lead to a divergence among the models by 2100.

In all of the scenarios, emissions reductions in electric power sector come at relatively lower prices than in other sectors (e.g. buildings, industry, and transport) so that the electricity sec-

tor is essentially decarbonized in the most stringent scenarios from all three modeling groups (Figure TS.14). At somewhat higher cost, other sectors can respond to rising carbon prices by reducing demands for fossil fuels, applying CCS technologies where possible, and substituting low-carbon energy sources such as bioenergy and low-carbon electricity or hydrogen. The amount of electricity used per unit of total primary energy increases in all of the stabilization scenarios (Figure TS.15), but those scenarios with the highest relative use of electricity tend to exhibit lower stabilization costs in part because of the larger role of decarbonized power generation. Assumptions regarding costs and performance of technologies to facilitate these adjustments, particularly in the post-2050 period, play an important role in determining stabilization costs.

The assumption of *when* flexibility links elements of each stabilization scenario through time. This in turn means that in addition to near-term technology availability, differences in assumptions about technology in the post-2050 period are reflected in near-term emissions reductions and GHG prices.

Figure TS.14. Percentage of World Electricity Production from Low- or Zero-Emissions Technologies Across Scenarios (percentage). All three modeling groups assumed sufficient technological options to allow for substantial reductions in carbon emissions from electricity production. Options include fossil power plants with CCS, nuclear power, and renewable energy such as hydroelectric power, wind power, and solar power. In all of the Level 1 scenarios, the electricity sector is almost fully decarbonized by the end of the century.

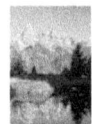

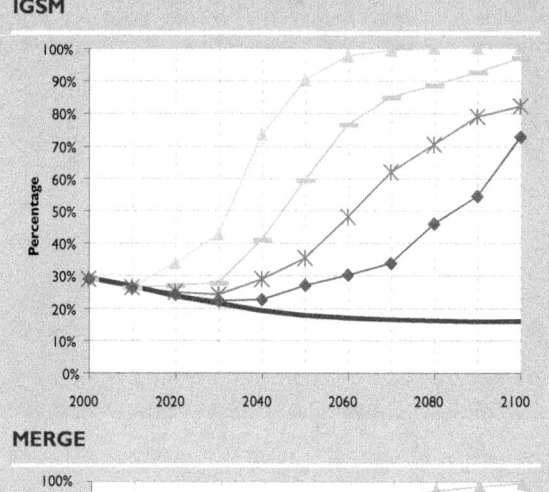

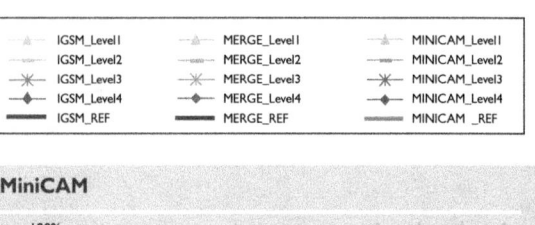

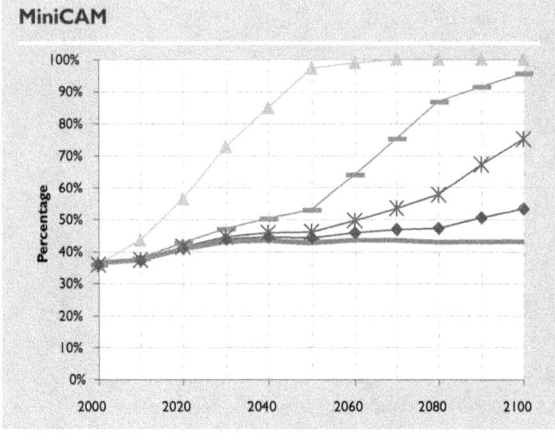

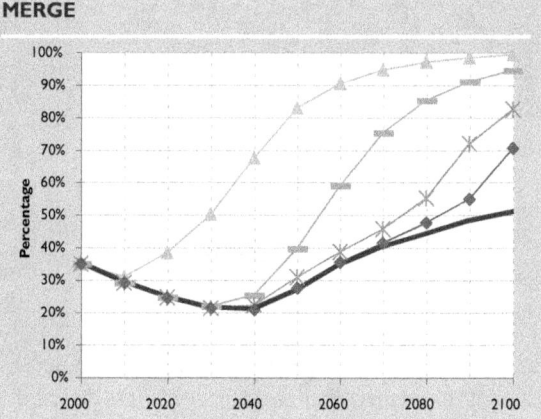

As noted earlier, the overall cost levels are strongly influenced by the idealized policy scenario that has all countries participating from the start, the assumption of *where* flexibility, an efficient pattern of emissions reductions over time, and integrated reductions in emissions of the different GHGs. Assumptions in which policies are implemented in a less efficient manner would lead to higher costs. Thus, these scenarios should not be interpreted as applying beyond the particular conditions assumed.

Constraints on GHG emissions also affect fuel prices. Generally, producer prices for fossil fuels fall as demand for them is depressed by the stabilization measures. Consumers of fossil fuels, on the other hand, pay for fuel plus a carbon price if the CO_2 emissions are freely re-

leased to the atmosphere (Table TS.5). Therefore, consumer costs of energy rise with more stringent stabilization levels in these scenarios.

Non-CO_2 gases play an important role in shaping the degree of change in the energy system. Scenarios that assume relatively better performance of technologies for reducing non-CO_2 emissions allow a given radiative forcing stabilization level to be met with greater radiative forcing from CO_2 and, all other things being equal, less extensive changes to the energy system. Differences in GHG concentrations among the three models reflect differences in assumed mitigation opportunities for non-CO_2 GHGs relative to CO_2. For example, lower CH_4 and N_2O emissions in the scenarios from one of the modeling groups reflects a greater market penetration of technologies that reduce CH_4 and N_2O emissions with positive profits even in the reference scenario, and significant abatement in the stabilization scenarios. With lower levels of CH_4 and N_2O than is the case in the scenarios

Figure TS.15. Ratio of Global Electricity Production to Primary Energy Consumption Across Scenarios.

Efforts to constrain CO_2 emissions result in increased use of electricity as a fraction of total primary energy in the scenarios from all three modeling groups. This is because all three modeling groups assumed lower-cost technology options for reductions in emissions from electricity production than for substitution away from fossil fuels in direct uses such as transportation. The scenarios from two of the modeling groups (MERGE and MiniCAM) generally include greater electrification than the scenarios from the third modeling group (IGSM). Greater opportunities to electrify reduce the economic impacts of stabilization. *[Note. Primary energy consumption from nuclear power and non-biomass renewable electricity are accounted for at the average efficiency of fossil-fired electric facilities, which vary over time and across scenarios.]*

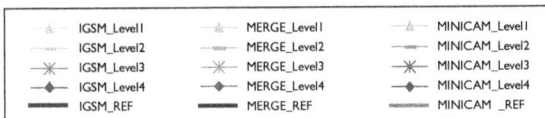

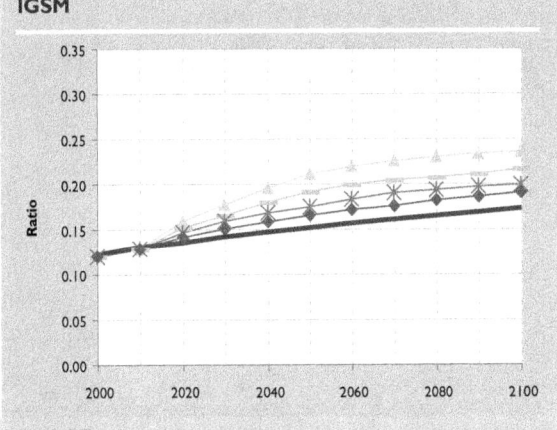

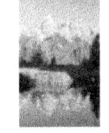

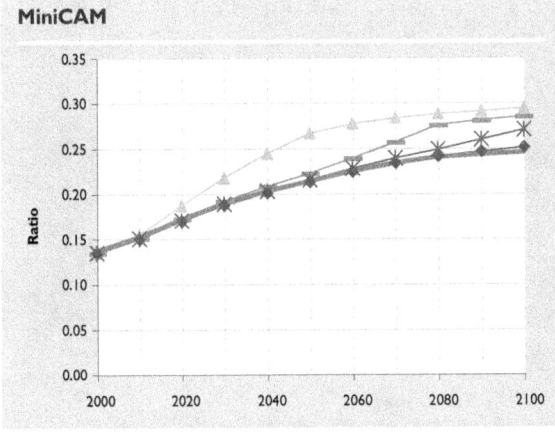

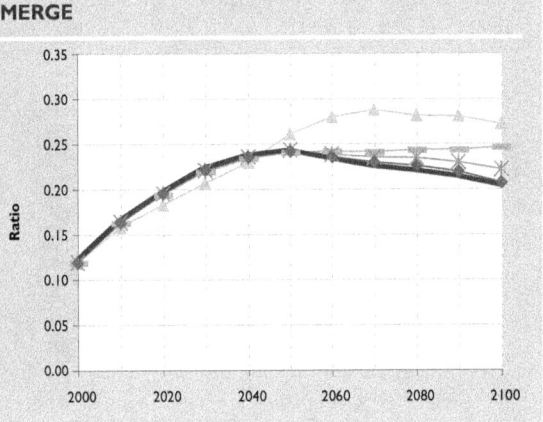

Table TS.5. Relationship Between a $100/tonne Carbon Price and Fuel Prices. (In most cases, stabilization depresses producer prices and so the percentage rise in the fuel cost seen by consumers would be less than indicated here. The change in producer price is highly scenario- and model-dependent.)

Fuel	Base Cost ($2005)	Added Cost ($)	Added Cost (%)
Crude Oil ($/bbl)	$60.0	$12.2	20%
Regular Gasoline ($/gal)	$2.39	$0.26	11%
Heating Oil ($/gal)	$2.34	$0.29	12%
Wellhead Natural Gas ($/tcf)	$10.17	$1.49	15%
Residential Natural Gas ($/tcf)	$15.30	$1.50	10%
Utility Coal ($/short ton)	$32.6	$55.3	170%
Electricity (c/kWh)	9.6¢	1.76¢	18%

Source: Bradley et al. 1991, updated with U.S. average prices for the 4th quarter of 2005 as reported in DOE 2006.

Figure TS.16. Total Radiative Forcing in 2100 Across Scenarios (W/m² from preindustrial). CO_2 is the main contributor to radiative forcing in the year 2100 in all of the scenarios. The opportunities to reduce control emissions from non-CO_2 GHGs influence the CO_2 emissions reductions required to meet a given radiative forcing stabilization level. At any stabilization level, scenarios with lower contributions to radiative forcing from non-CO_2 GHGs allow for greater radiative forcing from CO_2.

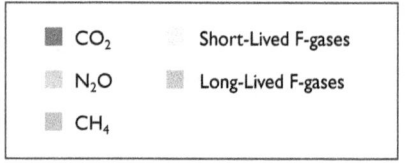

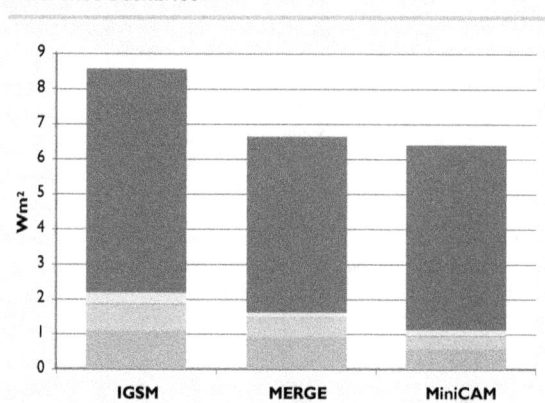

Reference Scenarios

Level 4 Scenarios

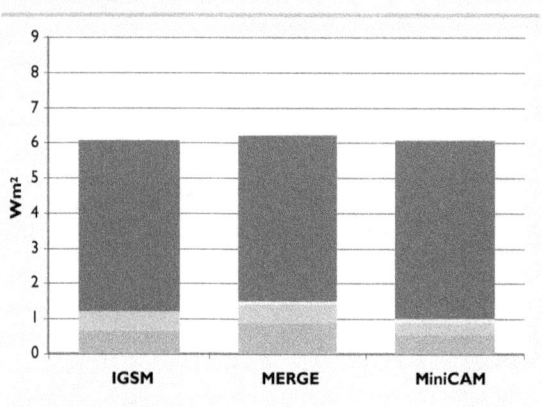

Level 3 Scenarios

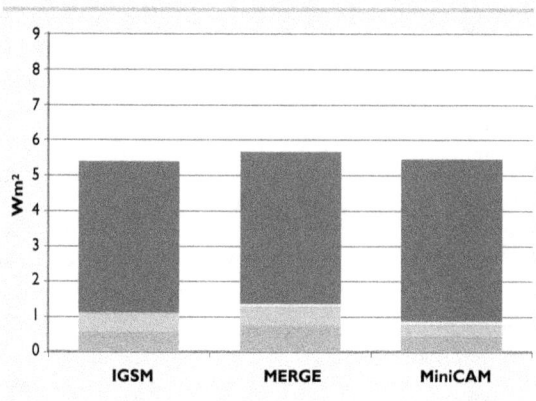

Level 2 Scenarios

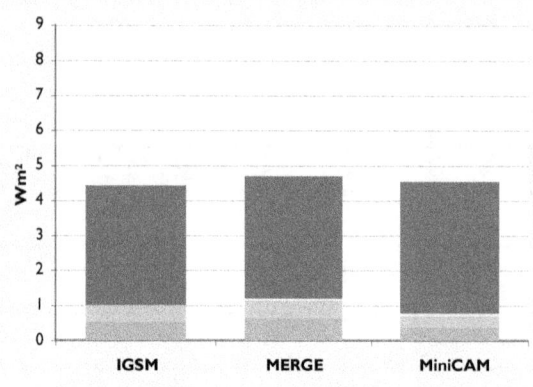

Level 1 Scenarios

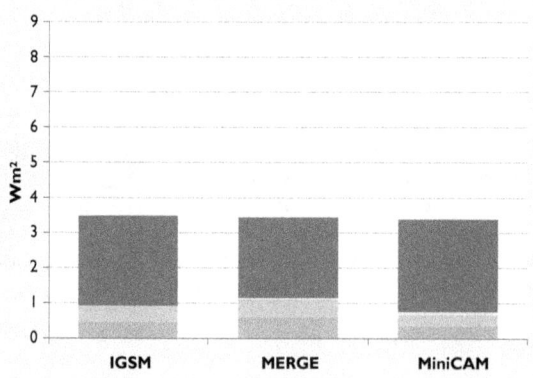

from the other two modeling groups, higher levels of CO_2 are still consistent with the overall radiative forcing levels (Figure TS. 16).

Achieving stabilization of atmospheric GHGs poses a substantial technological and policy challenge. It would require important transformations of the global energy system. The cost and feasibility of such a goal depends on the evolution of technology and its ability to overcome existing limits and barriers to adoption, and it depends on the efficiency and effectiveness of the policy instruments employed to achieve stabilization. These scenarios provide a means to gain insight into the challenge of stabilization and the implications of technology.

USING THE SCENARIOS AND FUTURE WORK

The scenarios in this report are intended as one of many inputs to public and private discussions regarding the threat of climate change, and they are also intended to serve as a point of departure for further CCSP and other analyses. A range of such analyses are possible. For example, the scenarios could be applied as the basis for assessing the climate implications of alternative stabilization levels, and then follow-on studies of potential climate impacts. They might also be used in studies exploring possible technology cost and performance goals, using information from the scenarios on energy prices and technology deployment levels. Similarly, the scenarios might inform analyses of the non-climate environmental implications of implementing potential new energy sources at a large scale. Another possibility is that the scenarios could serve as an input to a more complete analysis of the economic effects of stabilizing at the different radiative forcing levels, such as indicators of consumer impact in the U.S. (The reader is reminded, however, that these effects do not include the benefits that alternative stabilization levels might yield in reduced climate change risk or ancillary effects, such as effects on air pollution). The scenarios could also be compared against past and future scenarios analyses.

The scenarios in this report represent but one step in a long process of research and assessment, and the scenarios and their underlying models will benefit from further work. The review process has identified at least five different areas that hold the promise of potentially fruitful research: (1) technology sensitivity analysis, (2) consideration of non-idealized policy architectures, (3) expansion and improvement of the land use and terrestrial carbon cycle linkages to the energy and economic model components, (4) inclusion of other radiatively-important substances such as emissions affecting tropospheric ozone and aerosols, and (5) decision-making under uncertainty. These needs for additional research and analysis are elaborated in Chapter 5.

REFERENCES

CCSP, 2003: *Strategic Plan for the U.S. Climate Change Science Program*. A Report by the U.S. Climate Change Science Program and the Subcommittee on Global Change Research. Washington, D.C.

CCSP, 2005: *Final Prospectus for Synthesis and Assessment Product 2.1*. A Report by the U.S. Climate Change Science Program and the Subcommittee on Global Change Research. Washington, D.C.

CCSP, 2007: *Global Change Scenarios: Their Development and Use*. A Report by the U.S. Climate Change Science Program and the Subcommittee on Global Change Research [Parson, E., Burkett, V., Fisher-Vanden, K., Keith, D., Mearns, L., Pitcher, H., Rosenzweig, C., and Webster, M.]. U.S. Department of Energy, Washington, D.C.

U.S. DOE – Department of Energy, Energy Information Administration, 2006, *Short-Term Energy, and Winter Fuels Outlook* October 10th, 2006 Release

IPCC – Intergovernmental Panel on Climate Change. 2001. *Climate Change 2001: The Scientific Basis*. eds JT Houghton, Y Ding, DJ Griggs, N Noguer, PJ van der Linden, X Dai, K Maskell and CA Johnson, Cambridge University Press, Cambridge, U.K.

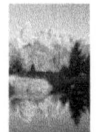

The U.S. Climate Change Science Program

Technical Summary

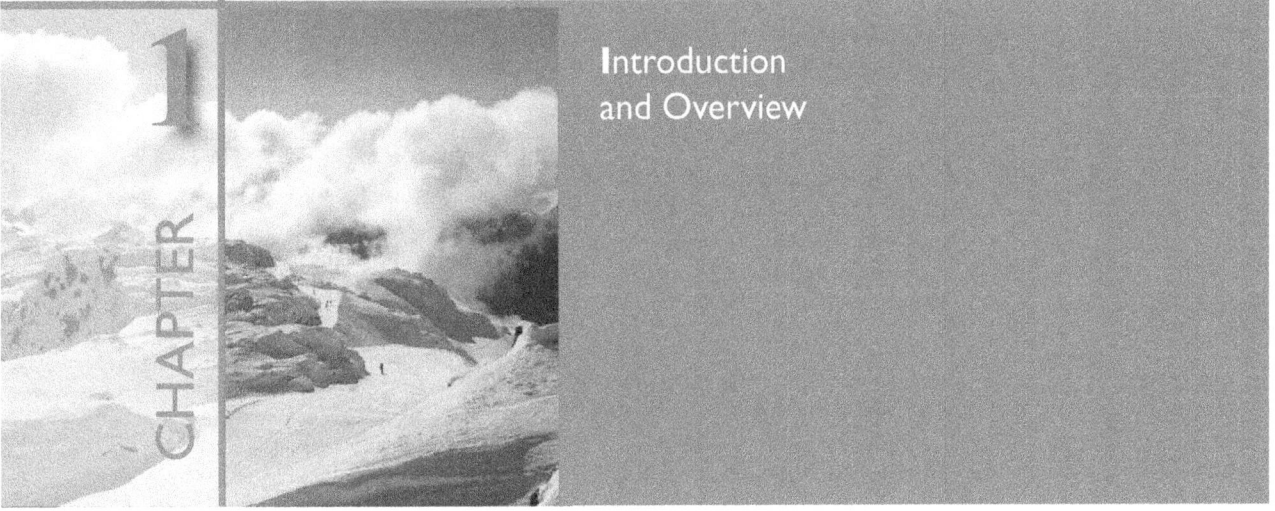

Introduction and Overview

INTRODUCTION

The Strategic Plan for the U.S. Climate Change Science Program (CCSP 2003) calls for the preparation of 21 synthesis and assessment products. Noting that "sound, comprehensive emissions scenarios are essential for comparative analysis of how climate might change in the future, as well as for analyses of mitigation and adaptation options," the Plan includes Product 2.1, *Scenarios of Greenhouse Gas Emissions and Atmospheric Concentrations and Review of Integrated Scenario Development and Application.* This report presents the scenarios created in the scenario-development component of Product 2.1; the review of scenario methods is the subject of a separate report (CCSP 2007). The guidelines for the development of these scenarios are set forth in the *Final Prospectus for Synthesis and Assessment Product 2.1* (CCSP 2005). Consistent with the Prospectus and the nature of the climate change issue, these scenarios were developed using long-term models of global energy-agriculture-land-use-economy systems coupled to models of global atmospheric composition and radiation.

This report discusses the overall design of scenarios (Chapter 1); describes the key features of the participating models (Chapter 2); presents and compares the newly prepared scenarios (Chapters 3 and 4); and discusses emerging insights from these new scenarios, the uses and limitations of the scenarios, and avenues for further research (Chapter 5). Scenario details are available in a separate data archive.

The scenarios in this report are intended as one of many inputs to public and private discussions regarding climate change and what to do about it, and they may also serve as a point of departure for further Climate Change Science Program (CCSP) and other analyses that might inform these discussions in the future. The possible users of these scenarios are many and diverse. They include climate modelers and the science community; those involved in national public policy formulation; managers of Federal research programs; state and local government officials who face decisions that might be affected by climate change and mitigation measures; and individual firms, non-governmental organizations, and members of the public. Such a varied clientele implies an equally diverse set of possible needs, and no single scenario research product can hope to fully satisfy all of these needs. The Prospectus for this research highlighted three particular areas in which the scenarios might provide valuable insights:

- *Emissions Trajectories.* What emissions trajectories over time are consistent with meeting the four stabilization levels, and what are the key factors that shape them?

- *Energy Systems.* What energy system characteristics are consistent with each of the four alternative stabilization levels, and how might these characteristics differ among stabilization levels?

- *Economic Implications.* What are the possible economic consequences of meeting each of the four alternative stabilization levels?

It should be emphasized that there are issues of climate change decision making that these scenarios do not address. For example, they were not designed for use in exploring the role of aerosols in climate change. Also, they lack the regional detail that may be desired for many aspects of local or regional decision-making.

> In addition, the scenarios in this report do not constitute a cost-benefit analysis of climate policy. They focus exclusively on the issues associated with reducing emissions to meet various stabilization levels; they do not consider the damages avoided through stabilization or ancillary benefits that could be realized by emissions reductions, such as reductions in local air pollution. Thus, although the scenarios should serve as a useful input to climate-related decision making, they address only one of several components of a benefit-cost analysis of climate policy.

Three analytical models, all meeting the criteria set forth in the Prospectus, were used in preparing the new scenarios. As also directed in the Prospectus, fifteen scenarios are presented in this document, five from each of the three modeling groups. First, each group produced a unique reference scenario based on the assumption that no climate policy would be implemented either nationally or globally beyond the current set of policies in place (e.g., the Kyoto Protocol and the President's greenhouse gas emissions intensity target for the U.S.). These reference scenarios were developed independently by the modeling groups, so they provide three separate visions of how the future

might unfold across the globe over the 21st century without additional climate policies.[1]

Each group then produced four additional stabilization scenarios, which are departures from each group's reference scenario. The Prospectus specified that stabilization levels, common across the groups, be defined in terms of the total long-term radiative impact of the suite of greenhouse gases (GHGs) that includes carbon dioxide (CO_2), nitrous oxide (N_2O), methane (CH_4), hydrofluorocarbons (HFCs), perfluorocarbons (PFCs), and sulfur hexafluoride (SF_6). This radiative impact is expressed in terms of radiative forcing associated with increases from preindustrial concentrations of this suite of GHGs (Box 1.1).

Although stabilization is defined in terms of radiative forcing, the stabilization levels were constructed so that the resulting CO_2 concentrations, after accounting for radiative forcing from the non-CO_2 GHGs, would be roughly 450 parts per million by volume (ppmv), 550 ppmv, 650 ppmv, and 750 ppmv. The radiative forcing limits therefore are higher than the forcing from CO_2 alone at these concentrations. Based on this requirement, the four stabilization levels were chosen as 3.4 watts per meter squared (W/m^2) (Level 1), 4.7 W/m^2 (Level 2), 5.8 W/m^2 (Level 3), and 6.7 W/m^2 (Level 4). In comparison, radiative forcing relative to preindustrial levels for this suite of gases stood at roughly 2.1 W/m^2 in 1998. Details of these stabilization assumptions are elaborated in Section 1.3 and Chapter 4.

The production of emissions scenarios consistent with these stabilization goals required analysis beyond the study of the emissions themselves because of physical, chemical, and biological feedbacks within the Earth system. Scenarios focused only on emissions of GHGs and other substances generated by human activity (anthropogenic sources) can rely exclusively on energy-agriculture-economic models that represent human activity and the emissions

---

[1]   Although there are many reasons to expect that the three reference scenarios would be different, it is worth noting that the modeling groups met periodically during the development of the scenarios to review progress and to exchange information. Thus, while not adhering to any formal protocol of standardization, the three reference scenarios are not entirely independent.

that result. However, relating emissions paths to concentrations of GHGs in the atmosphere requires models that account for both anthropogenic and natural sources as well as the sinks for these substances.

Models that attempt to capture these complex interactions and feedbacks must, because of computational limits, use simplified representations of individual components of the Earth system. These simplified representations are typically designed to mimic the behavior of more complex models but cannot represent all of the elements of these systems. Thus, while the scenario research undertaken here uses models that represent both the anthropogenic sources (the global energy-industrial-agricultural economy) and the Earth system processes (ocean, atmosphere, and terrestrial systems), it is not intended to supplant detailed analysis of these systems using full scale, state-of-the-art models and analytic techniques. Rather, these scenarios provide a common point of departure for more complex analyses of individual components of the Earth system as it is affected by human activity. These might include detailed studies of sub-components of the energy sector, regional scenarios of climate change using three-dimensional general circulation models (GCMs) and further downscaling techniques, and assessment of the implications of climate change under various stabilization goals for economic activity and natural ecosystems.

The remainder of this chapter is organized into four sections. Section 1.2 provides an overview of scientific aspects of the climate issue as background for interpretation of these scenarios. Section 1.3 then presents the research design with a focus on the characteristics of the stabilization scenarios to be investigated in Chapter 4. Section 1.4 briefly discusses how scenarios of this type have been used to examine the climate change issue and the intended uses and limits of the new scenarios, focusing on interpretation of these scenarios under conditions of uncertainty. Section 1.5 provides a guide to the structure of the remaining chapters.

BACKGROUND: HUMAN ACTIVITIES, EMISSIONS, CONCENTRATIONS, AND CLIMATE CHANGE

Materials that influence the Earth's radiation balance come in various forms, and most have natural as well as anthropogenic sources. Some are gases which remain in the atmosphere from days to millennia, trapping heat. They are known as GHGs because, while transparent to incoming short-wave radiation (the visible spectrum that people commonly perceive as light), they capture and reflect back to Earth long-wave radiation, thus increasing the temperature of the lower atmosphere. These naturally occurring GHGs, plus clouds and water vapor (the most important GHG of all), are responsible for creating a habitable climate on Earth. Without them, the average temperature at the Earth's surface would be colder than it is today by roughly 55°F (~30°C).

GHGs are not the only influences on the Earth's radiative balance. Other gases such as oxides of nitrogen (NO_x) have no direct greenhouse effect, but they are components of the atmospheric chemistry that determine the lifetime of some of the heat-trapping GHGs and are involved in the reactions that produce tropospheric ozone, another GHG. Aerosols (non-aqueous particles suspended in air) may have positive or negative effects, depending on their relative brightness. Some present a white surface and reflect the sun's energy back to space; others are black and absorb solar energy, adding to the solar warming of the atmosphere. Aerosols also have an indirect effect on climate in that they influence the character and lifetime of clouds, which have a strong influence on the radiation balance and on precipitation. Humans also alter the land surface, changing its reflective properties, and these changes can have climate consequences with effects most pronounced at a local scale (e.g., urban heat islands) and regional levels (e.g., large-scale changes in forest cover). In addition, the climate itself has positive and negative feedbacks, such as the decrease in global albedo that would result from melting land and sea ice or the potential release of GHGs, such as CH_4 from wetlands.

Climate policy concerns are driven by the fact that emissions from human activities (mainly combustion of fuels and biomass, industrial activities, and agriculture) are increasing the atmospheric concentrations of these substances. Climate policy discussions have focused heavily on CO_2, CH_4, N_2O, and a set of fluorine-containing industrial chemicals – SF_6 and two families of substances that do not exist naturally, hydrogenated halocarbons (including hydrochlorofluorocarbons [HCFCs] and HFCs)[2] and PFCs. Some of these substances remain in the atmosphere for decades (CH_4 and most HFCs), others for about 100 years (CO_2 and N_2O), and some for thousands of years (PFCs and SF_6).

Other naturally occurring substances whose levels have also been greatly enhanced by human activities remain in the atmosphere for days to months. With such short lifetimes, they are not well mixed in the atmosphere, so their effects have a regional pattern as well as global consequences. These substances include aerosols such as black carbon and other particulate matter; sulfur dioxide, which is the main precursor of the reflecting aerosols; and other gases such as volatile organic compounds, nitrogen dioxide, other oxides of nitrogen, and carbon monoxide. All are important components of atmospheric chemistry.

This suite of substances with different radiative potency and different lifetimes in the atmosphere presents a challenge in defining what is meant by atmospheric stabilization. Specification in terms of quantities of the substances themselves is problematic because there is no simple way to add them together in their natural units, such as tonnes or ppmv. Thus, a meaningful metric is needed to combine the effects of different GHGs.

One approach is to define stabilization in terms of some ultimate climate measure, such as the change in the global average temperature. One drawback of such measures is that they interject large uncertainties into the consideration of stabilization because the ultimate climate system

response to added GHGs is uncertain. Climate models involve complex and uncertain interactions and feedbacks, such as increasing levels of water vapor, changes in reflective polar ice, cloud effects of aerosols, and changes in ocean circulation that determine the ocean's uptake of CO_2 and heat.

For the design of these scenarios, the Prospectus called for an intermediate, less uncertain measure of climate effect. The Prospectus directed that stabilization "be defined in terms of the radiative forcing resulting from the long-term combined effects of carbon dioxide (CO_2), nitrous oxide (N_2O), methane (CH_4), hydrofluorocarbons (HFCs), perfluorocarbons (PFCs), and sulfur hexafluoride (SF_6)." Radiative forcing (Box 1.1) is a measure of the instantaneous imbalance in the radiative energy budget of the Earth's climate system (energy in versus energy out) resulting from an externally imposed perturbation such as increasing GHG concentrations. It is measured in terms of W/m^2 at the Earth's shell and a positive value means a warming influence. For these scenarios, radiative forcing is measured against the concentrations of the GHGs considered in this research in preindustrial times, taken to be 1750.

Figure 1.1 shows estimates of how increases in GHGs, aerosols, and other changes have influenced radiative forcing since 1850. The GHGs considered in these scenarios are collected in the left-most bar and together they have had the biggest effect, with CO_2 being the largest of this group. Increased tropospheric ozone has also had a substantial warming effect. The reduction in stratospheric ozone has had a slight cooling effect. Changes in aerosols have had both warming and cooling effects. Aerosol effects are highly uncertain because they depend on the nature of the particles; how the particles are distributed in the atmosphere; and the concentrations of the particles, which are not as well understood as the GHGs. Land-use change and its effect on the reflectivity of the Earth's surface, jet contrails and changes in high-level (cirrus) clouds, and the natural change in intensity of the sun have also had effects.

Another important aspect of the climate effects of these substances, not captured in the W/m^2

---

2   For simplicity, all hydrogenated halocarbons will be referred to as HFCs in the subsequent text. The greenhouse gas methyl chloroform is often also grouped along with HFCs and HCFCs.

BOX 1.1 Radiative Forcing

Most of the Sun's energy that reaches the Earth is absorbed by the oceans and land masses and radiated back into the atmosphere in the form of heat or infrared radiation. Some of this infrared energy is absorbed and re-radiated back to the Earth by atmospheric gases, including water vapor, CO_2, and other substances. As concentrations of GHGs increase, there are direct and indirect effects on the Earth's energy balance. The direct effect is often referred to as a radiative forcing, a subset of a more general set of phenomena referred to as climate forcings. The National Research Council (NRC 2005) offers the following set of definitions:

> Factors that affect climate change are usefully separated into forcings and feedbacks.... A climate forcing is an energy imbalance imposed on the climate system either externally or by human activities. Examples include changes in solar energy output, volcanic emissions, deliberate land modification, or anthropogenic emissions of greenhouse gases, aerosols, and their precursors. A climate feedback is an internal climate process that amplifies or dampens the climate response to an initial forcing. An example is the increase in atmospheric water vapor that is triggered by an initial warming due to rising carbon dioxide (CO_2) concentrations, which then acts to amplify the warming through the greenhouse properties of water vapor....

> Climate forcing: An energy imbalance imposed on the climate system either externally or by human activities.

> • **Direct radiative forcing:** A climate forcing that directly affects the radiative budget of the Earth's climate system; for example, added carbon dioxide (CO_2) absorbs and emits infrared radiation. Direct radiative forcing may be due to a change in concentration of radiatively active gases, a change in solar radiation reaching the Earth, or changes in surface albedo. Radiative forcing is reported in the climate change scientific literature as a change in energy flux at the tropopause, calculated in units of watts per square meter (W/m^2); model calculations typically report values in which the stratosphere was allowed to adjust thermally to the forcing under an assumption of fixed stratospheric dynamics.

> • **Indirect radiative forcing:** A climate forcing that creates a radiative imbalance by first altering climate system components (e.g., precipitation efficiency of clouds), which then almost immediately lead to changes in radiative fluxes. Examples include the effect of solar variability on stratospheric ozone and the modification of cloud properties by aerosols.

> • **Nonradiative forcing:** A climate forcing that creates an energy imbalance that does not immediately involve radiation. An example is the increasing evapotranspiration flux resulting from agricultural irrigation.

For purposes of this report, the radiative forcing stabilization levels are defined in terms of the direct radiative forcing caused by increases from preindustrial concentrations of CO_2, CH_4, N_2O, PFCs, HFCs, and SF_6. The indirect radiative effects are not included in calculating whether the radiative forcing stabilization level levels are met, nor are the direct radiative effects (positive or negative) of other substances such as ozone, CFCs, or aerosols, although emissions of these substances and their radiative and climatic effects are part of these integrated system models.

measure, is the persistence of their influence on the radiative balance – a characteristic discussed in Box 1.2. The W/m^2 measure of radiative forcing accounts for only the effect of a concentration in the atmosphere at a particular instant. The GHGs considered here have influences that may last from a decade or two (e.g., the influence of CH_4) to millennia, as noted earlier.

An important difference between GHGs and most of the other substances in Figure 1.1 is their long lifetimes. In contrast to GHGs,

> ## BOX 1.2 Atmospheric Lifetimes of Greenhouse Gases
>
> The atmospheric lifetime concept is more appropriate for CH_4, N_2O, HFCs, PFCs, and SF_6 than it is for CO_2. These non-CO_2 gases are destroyed via chemical processes after some time in the atmosphere. In contrast, CO_2 is constantly cycled between pools in the atmosphere, the surface layer of the ocean, and vegetation, so it is (for the most part) not destroyed. Very slow processes lead to some removal of carbon from oceans, vegetation, and the atmosphere as calcium carbonate. Also, over long geological periods, carbon from vegetation is stored as fossil fuels, which is a permanent removal process as long as the fossil fuels are not burned to produce energy.
>
> Although the lifetime concept is not strictly appropriate for CO_2 (see Box 2.2 in Chapter 2), the molecules in a kilogram of emissions can be thought of as residing in the atmosphere, exercising their radiative effect, for around 100 years. This approximation allows a rough comparison with the other gases: CH_4 at 12 years, N_2O at 114 years, and SF_6 at 3200 years. HFCs are a family of gases with varying lifetimes from less than a year to over 200 years; those predominantly in use now have lifetimes mostly in the range of 10 to 50 years. Similarly, the PFCs have various lifetimes, ranging from 2,600 to 50,000 years.
>
> The lifetimes are not constant, as they depend to some degree on other Earth system processes. The lifetime of CH_4 is the most affected by the levels of other pollutants in the atmosphere.

aerosols remain in the atmosphere only for a few days to a couple of weeks. Once an aerosol emission source is eliminated, its effect on radiative forcing disappears very quickly. Tropospheric ozone lasts for a few months. Moreover, relatively short-lived substances are not well mixed in the atmosphere. Levels are very high near emissions sources and much lower in other parts of the world, so their climate effect has a different spatial pattern than that of long-lived substances. The regional differences and much shorter lifetimes of non-GHG substances make comparisons among them more difficult than among GHGs. The radiative effects of these substances also subject to more uncertainty, as shown in Figure 1.1.

RESEARCH DESIGN

The broad elements of the research design for these scenarios are set forth in the Prospectus, including (1) selection of models, (2) guidance to the modeling groups for development of a reference scenario, and (3) guidance for the development of stabilization scenarios.

Figure 1.1. Estimated Influences of Atmospheric Gases on Radiative Forcing, 1750-Present. Source: IPCC 2001

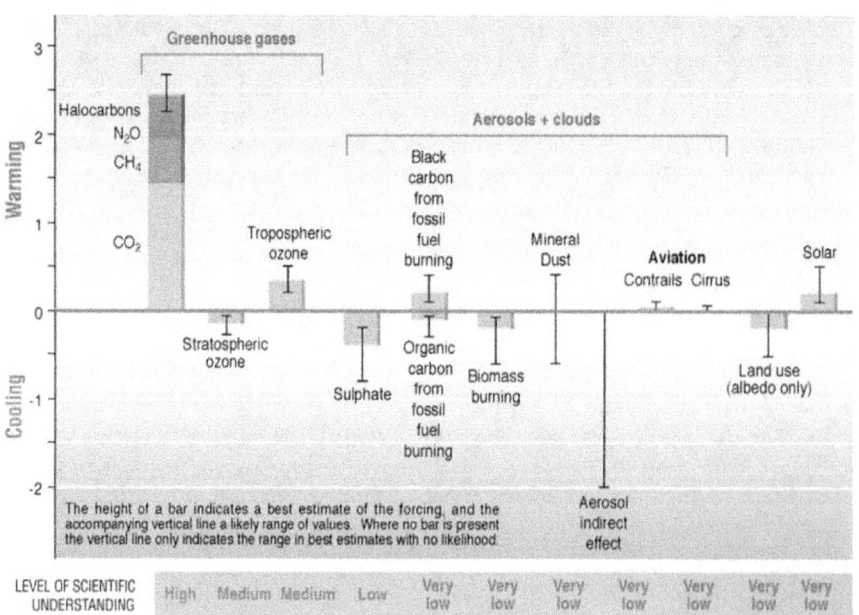

Model Selection

The Prospectus set forth the model capabilities required to develop the desired stabilization scenarios. As stated in the Prospectus, participating models must:

1. Be global in scale

2. Be capable of producing global emissions totals for, at a minimum, CO_2, N_2O, CH_4, HFCs, PFCs, and SF_6 that may serve as inputs to global GCMs, such as the National Center for Atmospheric Research (NCAR) Community Climate System Model and the Geophysical Fluid Dynamics Laboratory climate model

3. Be capable of simulating the radiative forcing from CO_2, N_2O, CH_4, HFCs, PFCs, and SF_6

4. Represent multiple regions

5. Have technological resolution capable of distinguishing among major sources of primary energy (e.g., renewable energy, nuclear energy, biomass, oil, coal, and natural gas) as well as between fossil fuel technologies with and without CO_2 capture and storage (CCS) systems

6. Be economics based and capable of simulating macroeconomic cost implications of stabilization

7. Look forward to the end of the century or beyond.

In addition, the Prospectus required that the modeling groups have a track record of publications in professional, refereed journals, specifically in the use of their models for the analysis of long-term GHG emission scenarios.

Selection by these criteria led to the three models used in this research: (1) The Integrated Global Systems Model (IGSM) of the Massachusetts Institute of Technology's Joint Program on the Science and Policy of Global Change; (2) the Model for Evaluating the Regional and Global Effects of GHG reduction policies (MERGE), developed jointly at Stanford University and the Electric Power Research Institute; and (3) the MiniCAM Model of the Joint Global Change Research Institute, which is a partner-

ship between the Pacific Northwest National Laboratory and the University of Maryland.

Each of these models has been used extensively for climate change analysis. The roots of each extend back more than a decade, during which time features and details have been added. Analyses using each have appeared widely in peer-reviewed publications. The features of the models are described in Chapter 2 with references to publications and reports that provide complete documentation.

These models fall into a class that has come to be known as Integrated Assessment Models (IAMs). There are many ways to define IAMs and to characterize the motivations for developing them (IPCC 1996). A particularly appropriate definition of their primary purposes, provided by Parson and Fisher-Vanden (1997), is "evaluating potential responses to climate change, structuring knowledge and characterizing uncertainty, contributing to broad comparative risk assessments, and contributing to scientific research."

Development of Reference Scenarios

As required by the Prospectus, each participating modeling group first produced a reference scenario that assumes no policies specifically intended to address climate change beyond implementation of any existing policies to the end of their commitment periods, including the Kyoto Protocol and the policy of the U.S. to reduce greenhouse gas emissions intensity by 18% by 2012. For purposes of the reference scenario (and for each of the stabilization scenarios), it was assumed that these policies are successfully implemented through 2012 and their goals are achieved. (This assumption could only be approximated within the models because their time steps did not coincide exactly with the period from 2002 to 2012. However, such approximation is a minor consideration as slight differences in emissions for a few years have little impact on long term concentrations.) As directed by the Prospectus, after 2012 these existing climate policies expire and are not renewed or replaced. This is not a prediction or a best-judgment forecast, but a scenario designed to provide a clearly defined point of departure

for illuminating the implications of alternative stabilization goals. The paths toward stabilization are implemented to start after 2012 as discussed further in the following section. The reference scenarios and assumptions underlying them are detailed in Chapter 3.

The reference scenarios serve two main purposes. First, they provide insight into how the world might evolve without additional efforts to constrain GHG emissions, given various assumptions about principal drivers of the economy, energy use, and emissions. These assumptions include those concerning population increase, land and labor productivity growth, technological options, and resource endowments. These forces govern the supply and demand for energy, industrial goods, and agricultural products – the production and consumption activities that lead to GHG emissions. The reference scenarios are a thought experiment in that they assume that even as emissions increase and climate changes nothing is done to reduce emissions. The specific levels of GHG emissions and concentrations are not predetermined but result from the combination of assumptions made.

Second, the reference scenarios serve as points of departure for analysis of the changes brought about by stabilization of radiative forcing, and the underlying assumptions have a large bearing on the characteristics of the stabilization scenarios. For example, all other things being equal, the lower the economic growth and the higher the availability and competitiveness of low-carbon energy technologies in the reference scenario, the lower will be the GHG emissions and the easier it will be to reach stabilization. On the other hand, if a reference scenario assumes that fossil fuels are abundant, and fossil fuel technologies will become cheaper over time while low- or zero-carbon alternatives remain expensive, the scenario will show consumers having little reason to conserve, adopt more efficient energy equipment, or switch to non-fossil sources. Under such a reference scenario, emissions will grow rapidly, and stronger economic incentives will be required to achieve stabilization.

Finally, the Prospectus specified that the modeling groups develop their reference scenarios independently[3], applying *meaningful* and *plausible* assumptions for key drivers. Similarities and differences among the reference scenarios are useful in illustrating the uncertainty inherent in long-run treatment of the climate challenge. At the same time, with only three participating models, the range of scenario assumptions produced does not span the full range of possibilities.

Development of the Stabilization Scenarios

Although the model groups were required to independently develop their modeling assumptions, the Prospectus specified that a common set of four stabilization targets be used across the participating models. Also, whereas much of the literature on atmospheric stabilization focuses on concentrations of CO_2 only, an important objective of this research was to expand the range of coverage to include other GHGs. Thus, the Prospectus required that the stabilization levels be defined in terms of the combined effects of CO_2, N_2O, CH_4, HFCs, PFCs, and SF_6. This suite of GHGs forms the basis for the U.S. GHG-intensity-reduction policy, announced by the President on February 14, 2002; it is the same set subject to control under the Kyoto Protocol. These gases are included in the left-most bar of Figure 1.1. The stabilization targets specified in the Prospectus explicitly omit the aerosol, ozone, land surface, and other effects shown in Figure 1.1, which may be influenced by measures taken to achieve the stabilization goal. Table 1.1 shows the change in concentration levels for these gases from 1750 to 2000. The left-most bar in Figure 1.1 shows radiative forcing of roughly 2.4 W/m^2 compared with a sum of 2.1 W/m^2 in Table 1.1. The difference exists because Figure 1.1 includes roughly 0.3 W/m^2 of forcing from chlorofluorocarbons (CFCs) not in Table 1.1. CFCs, important in the historical data, are already being phased out under the Montreal Protocol because of their stratospheric ozone-depleting properties, so they are not expected to be a significant source of additional increased forcing in the future. The HFCs, which do not contribute to stratospheric ozone depletion, were developed as substitutes

---

[3]   See footnote 1.

	Preindustrial Concentration (1750)	Current Concentration (1998)	Contribution to Radiative Forcing, (W/m², 1750 to 1998)
CO₂	278 ppmv	365 ppmv	1.46
CH₄	700 ppbv	1745 ppbv	0.48
N₂O	270 ppbv	314 ppbv	0.15
HFCs, PFCs, SF₆	0	various	≈ 0.02
Total	—	—	≈ 2.1
Source: IPCC 2001.			

Table 1.1. Greenhouse Gas Concentrations and Forcing.

Concentrations of GHGs have increased since 1750 (preindustrial), altering the radiative energy budget of the Earth's climate system.

for the CFCs, but are of concern because of their radiative properties. Table 1.2 shows the specific radiative forcing targets chosen.

As noted earlier, the Prospectus instructed that the stabilization levels be constructed so that the CO₂ concentrations resulting from stabilization of total radiative forcing, after accounting for radiative forcing from the non-CO₂ GHGs, would be roughly 450 ppmv, 550 ppmv, 650 ppmv, and 750 ppmv. This correspondence was achieved by (1) calculating the increased radiative forcing from CO₂ at each of these concentrations, (2) adding to that amount the radiative forcing from the non-CO₂ gases from 1750 to present, and (3) adding an estimate of the change in radiative forcing from the non-CO₂ GHGs under each of the stabilization levels. Each of the models represents the emissions and abatement opportunities of the non-CO₂ gases somewhat differently and takes a different approach to representation of the tradeoffs among them, so an exact correspondence between overall radiative forcing and CO₂ levels that would fit all three models was not possible.

The Prospectus also specified that, beyond the implementation of any existing policies, the stabilization scenarios should be based on universal participation by the world's nations. This guidance was implemented by assuming a climate regime with simultaneous global participation in emissions mitigation and in which the marginal costs of emission controls are equalized across countries and regions. Under this assumption, known as *where* flexibility, emissions will be reduced where it is cheapest to do so regardless of their geographical location. One important implication of this assumption is that the stabilization scenarios produce estimates of stabilization costs that are systematically lower than what might be expected in a world in which some major countries remain out of an emissions mitigation regime for an extended period of time, some economies use more costly regulatory mechanisms, or emissions mitigation regimes within nations are incomplete either in terms of GHG or sectoral coverage. On the other hand, possible ancillary benefits, tax interaction effects, or effects of carbon policies on technical change were not considered, which in

	Total Radiative Forcing from GHGs (W/m²)	Approximate Contribution to Radiative Forcing from non-CO₂ GHGs (W/m²)	Approximate Contribution to Radiative Forcing from CO₂ (W/m²)	Corresponding CO₂ Concentration (ppmv)
Level 1	3.4	0.8	2.6	450
Level 2	4.7	1.0	3.7	550
Level 3	5.8	1.3	4.5	650
Level 4	6.7	1.4	5.3	750
Year 1998	≈ 2.1	0.65	1.46	365
Preindustrial (1750)	—	—	—	278

Table 1.2. Radiative Forcing Stabilization Levels (W/m²) and Approximate CO₂ Concentrations (ppmv). The radiative forcing levels were constructed so that the CO₂ concentrations resulting from stabilization of total radiative forcing, after accounting for radiative forcing from the non-CO₂ GHGs, would be roughly 450 ppmv, 550 ppmv, 650 ppmv, and 750 ppmv.

some cases can lower costs. These issues are discussed in more detail in Chapter 4.

In addition, the Prospectus required that stabilization be defined as long term. Because of the inertia in the Earth system, largely attributable to the ocean, perturbations to the climate and atmosphere have effects for thousands of years. Economic models have little credibility over such timeframes. The Prospectus, therefore, instructed that the participating modeling groups report scenario information only up through 2100. Each group then had to address how to relate the level in 2100 to the long-term goal. The chosen approaches were generally similar, but with some differences in implementation. This and other details of the stabilization scenario design are addressed more completely in Chapter 4.

INTERPRETING SCENARIOS: USES, LIMITS, AND UNCERTAINTY

Emissions scenarios have proven to be useful aids to understanding climate change, and there is a long history of their use (see Box 1.3). Scenarios are descriptions of future conditions, often constructed by asking *what if* questions, such as what if events were to unfold in a particular way? Informal scenario analysis is part of almost all decision making. For example, families making decisions about big purchases, such as a car or a house, might plausibly construct a scenario in which changes in employment forces them to move. Scenarios addressing major public-policy questions perform the same purpose, helping decision makers and the public to understand the consequences of actions today in the light of plausible future developments.

Models assist in creating scenarios by showing how assumptions about key drivers, such as economic and population growth or policy op-

---

BOX 1.3 Emissions Scenarios and Climate Change

Emissions scenarios that describe future economic growth and energy use have been important tools for understanding the long-term consequences of climate change. They were used in assessments by the U.S. National Academy of Sciences in 1983 and by the Department of Energy in 1985 (NAS 1983, US DOE 1985). Previous emissions scenarios have evolved from simple projections that extrapolated a 1% per year increase in CO_2 emissions to scenarios that incorporate assumptions about population, economic growth, energy supply, and controls on GHG emissions and CFCs (Leggett et al. 1992, Pepper et al. 1992). They played an important role in the reports of the Intergovernmental Panel on Climate Change (IPCC 1991, IPCC 1992, IPCC 1996). The IPCC Special Report on Emissions Scenarios (SRES) (Nakicenovic et al. 2000) was the most recent major effort undertaken by the IPCC to expand and update earlier scenarios. This set of scenarios was based on storylines of alternative futures, updated with regard to the variables used in previous scenarios and with additional detail on technological change and land use.

The Energy Modeling Forum (EMF) has been an important venue for intercomparison of emissions scenarios and IAMs. The EMF, managed at Stanford University, includes participants from academic, government, and other modeling groups from around the world. It has served this role for the energy-modeling community since the 1970s. Individual EMF studies run over a course of about two years, with scenarios designed by the participants to provide insight into the behavior of the participating models. Scenarios are often published in the peer-reviewed literature. A recent study, EMF 21, focused on multi-gas stabilization scenarios (de la Chesnaye and Weyant 2006).

tions, lead to particular levels of GHG emissions. Model-based scenario analysis is designed to provide quantitative estimates of multiple outcomes and to assure consistency among them that is difficult to achieve without a formal structure. Thus, a main benefit of such model simulation of scenarios is that they ensure basic accounting identities: the quantity demanded of fuel is equal to the quantity supplied, imports in one region are balanced by exports from other regions, cumulative fuel used does not exceed estimates of the available resources, and expenditures for goods and services do not exceed income. The approach complements other ways of thinking about the future, ranging from formal uncertainty analysis to narratives. Also, such model analyses offer a set of macroscenarios that users can build on, adding more detailed assumptions about variables and decisions of interest to them.

The possible users of these scenarios are many and diverse, and a single scenario research product cannot hope to provide the details needed by all potential users or to address their specific questions. Thus, these scenarios are an initial set offered to potential user communities. If successful, they will generate further questions and the demand for more detailed analysis, some of which might be satisfied by further scenario development from models like those used here, but more often demanding detail that can only be provided with other modeling and analysis techniques. As such, this effort is one step in an ongoing and iterative process of producing and refining climate-related scenarios and scenario tools.

Although the required long-term perspective demands scenarios that stretch into the distant future, any such scenarios carry with them considerable uncertainty. Inevitably, the future will hold surprises. Scientific advances will be made, new technologies will be developed, and the direction of the economy will change, making it necessary to reassess the issues examined here. The Prospectus called for development of a limited number of scenarios, without a formal treatment of likelihood or uncertainty, requiring as noted earlier, only that the modeling groups use assumptions that they believe to be *mean-*

ingful and *plausible*. Formal uncertainty analysis has much to offer and could be a useful additional follow-on or complementary research task. Here, however, the range of outcomes from the different modeling groups help to illustrate, if incompletely, the range of possibilities.

The scenarios developed here take the best information available now and assess what it may mean for the future. Any such research, however, will necessarily be incomplete and will not foresee all possible future developments. The best planning must prepare for changes in course later as new information becomes available.

REPORT OUTLINE

Chapter 2 of this report provides an overview of the three models used in development of the scenarios. Chapter 3 describes the assumptions about key drivers in each of the models and reports the reference scenarios. Chapter 4 provides greater detail on the design of the stabilization scenarios and then presents these scenarios. Chapter 5 provides concluding observations, including possible avenues for additional research.

The chapters seek to show how the models and the assumptions used by the modeling groups to develop the scenarios differ and, to the degree possible, to relate where these differences matter and how they shape the scenarios. The models have their own respective areas of focus, and each offers its own reasonable representation of the world. The authors have distilled general conclusions common to the scenarios generated by the three modeling groups, while recognizing that other plausible representations could well lead to quite different scenarios. The scenarios are presented primarily in the figures. Associated with the report is a database with quantitative information available for those who wish to further analyze and use these scenarios. A description of the database, directions for use, and its location can be found in the appendix.

Models Used
in This Research

OVERVIEW OF THE MODELS

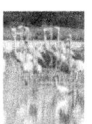

The computer models used in this research are referred to as integrated assessment models because they combine, in an integrated framework, the socioeconomic and physical processes and systems that define the human influence on, and interactions with, the global climate. They integrate computer models of socioeconomic and technological determinants of the emissions of GHGs and other substances influencing the Earth's radiation balance with models of the natural science of Earth system response, including those of the atmosphere, oceans, and terrestrial biosphere. Although they differ in their specific design objectives and details of their mathematical structures, each of these IAMs was developed for the purpose of gaining insight into economic and policy issues associated with global climate change.

To create scenarios of sufficient depth, scope, and detail, a number of model characteristics were deemed critical for development of these scenarios. The criteria set forth in the Prospectus for this research led to the selection of three IAMs: IGSM, MERGE, and MiniCAM. These three are among the most detailed models of this type of IAM, and each a has long history of development and application.

- **IGSM** of the Massachusetts Institute of Technology's Joint Program on the Science and Policy of Global Change is an Earth system model that comprises a multi-sector, multi-region economic component and a science component, including a two-dimensional atmosphere, a three-dimensional ocean, and a detailed biogeochemical model of the terrestrial biosphere (Sokolov et al. 2005). Because this research focuses on new emissions scenarios, elements of the scenarios emerging from the economic model component of IGSM, the Emissions Prediction and Policy Analysis (EPPA) model (Paltsev et al. 2005), are featured in the discussion below. EPPA is a recursive-dynamic computable general equilibrium (CGE) model of the world economy and greenhouse-relevant emissions, solved on a five-year time step. Previous applications of IGSM and its EPPA component system can be found at http://web.mit.edu/globalchange.

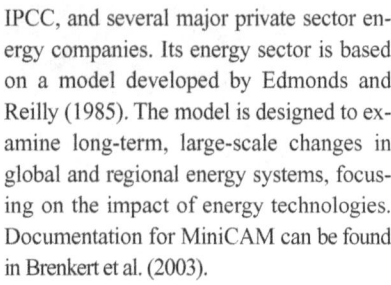

- **MERGE** was developed jointly at Stanford University and the Electric Power Research Institute (Manne and Richels 2005). It is an inter-temporal general equilibrium model of the global economy in which the world is divided into nine geopolitical regions. It is solved on a ten-year time step. MERGE is a hybrid model, combining a bottom-up representation of the energy supply sector with a top-down perspective on the remainder of the economy.[1] Savings and investment decisions are modeled as if each region maximizes the discounted utility of its consumption, subject to an inter-temporal wealth constraint. Embedded within this structure is a reduced-form representation of the physical Earth system. MERGE has been used to explore a range of climate-related issues, including multi-gas strategies, the value of low-carbon-emitting energy technologies, the choice of near-term hedging strategies under uncertainty, the impacts of learning-by-doing, and the potential importance of *when* and *where* flexibility. To support this scenario research, the multi-gas version has been revised by adjustments in technology and other assumptions. The MERGE code and publications describing its structure and applications can be found at http://www.stanford.edu/group/MERGE/.

- **MiniCAM** is an integrated assessment model (Brenkert et al. 2003) that combines a technologically detailed global energy-economy-agricultural-land-use model with a suite of coupled gas-cycle, climate, and ice-melt models, integrated in the Model for the Assessment of Greenhouse-Gas Induced Climate Change (MAGICC). MiniCAM was developed and is maintained at the Joint Global Change Research Institute, a partnership between the Pacific Northwest National Laboratory and the University of Maryland, while MAGICC was developed and is maintained at NCAR. MiniCAM is solved on a 15-year time step. MiniCAM has been used extensively for energy, climate, and other environmental analyses conducted for organizations that include the U.S. Department of Energy (DOE), the U.S. Environmental Protection Agency, the

IPCC, and several major private sector energy companies. Its energy sector is based on a model developed by Edmonds and Reilly (1985). The model is designed to examine long-term, large-scale changes in global and regional energy systems, focusing on the impact of energy technologies. Documentation for MiniCAM can be found in Brenkert et al. (2003).

Because these models were designed to address an overlapping set of climate change issues, they are similar in many respects. All three have social science-based components that capture the socioeconomic and technology interactions underlying the emissions of GHGs, and each incorporates models of physical cycles for GHGs and other radiatively important substances and other aspects of the natural science of global climate. The differences among them lie in the detail and construction of these components and in the ways they are modeled to interact. Each was designed with somewhat different aspects of the climate issue as a main focus. IGSM includes the most detailed representation of the chemistry, physics, and biology of the atmosphere, oceans, and terrestrial biosphere; thus, its EPPA component is designed to provide the emissions detail that these natural science components require. MERGE has its origins in an energy-sector model that was initially designed for energy technology assessment. It was subsequently modified to explore the influence of expectations (and uncertainty regarding expectations) about future climate policy on the economics of current investment and the cost-minimizing allocation of emissions mitigation over time. Its focus requires a forward-looking structure, which in turn employs simplified non-energy components of the economy. MiniCAM is a technology-rich IAM. It features detailed representations of energy technologies, energy systems, and energy markets and their interactions with demographics, the economy, agricultural technologies, markets, land use, and the terrestrial carbon cycle.

Each of these IAMs has unique strengths and areas of special insight. In this research, the simultaneous application of different model structures is useful in revealing different aspects of the task of stabilizing radiative forcing. The differences among the scenarios prepared by the

---

[1] It differs from the pure bottom-up approach described in Box 2.1 in that demands for energy are price responsive.

three modeling groups, presented in Chapters 3 and 4, are an indication of the limits of the knowledge about future GHG emissions and the challenges in stabilizing atmospheric conditions. Indeed, differences among the emissions characteristics of the reference scenarios and in the implications of various stabilization targets are likely within the range that would be realized from an uncertainty analysis applied to any one of the three, as indicated by the analysis of the IGSM model by Webster et al. (2003).

Table 2.1 provides a cross-model overview of some of the key characteristics to be compared in the following sections of this chapter. Section 2.2 focuses on social science components, describing similarities and differences and highlighting the assumptions that have the greatest influences on the scenarios. Section 2.3 does the same for the natural science sub-models of each IAM, which in this research make the connection between the emissions of GHGs and the resulting atmospheric conditions.

Table 2.1. Characteristics of the Models

Feature	IGSM (with EPPA Economics Component)	MERGE	MiniCAM
Regions	16	9	14
Time Horizon, Time Steps	2100, 5-year steps	2200, 10-year steps	2095, 15-year steps
Model Structure	General equilibrium	General equilibrium	Partial equilibrium
Solution	Recursive dynamic	Inter-temporal optimization	Recursive dynamic
Final Energy Demand Sectors in Each Region	Households, private transportation, commercial transportation, service sector, agriculture, energy intensive industries, and other industry	A single, non-energy production sector	Buildings, transportation, and industry (including agriculture)
Capital Turnover	Five vintages of capital with a depreciation rate	A putty clay approach wherein the input-output coefficients for each cohort are optimally adjusted to the future trajectory of prices at the time of investment	Vintages with constant depreciation rate for all electricity-sector capital; capital structure not explicitly modeled in other sectors
Goods in International Trade	All energy and non-energy goods as well as emissions permits	Energy, energy intensive industry goods, emissions permits, and representative tradable goods	Oil, coal, natural gas, biomass, agricultural goods, and emissions permits
Emissions	CO_2, CH4, N_2O, HFCs, PFCs, SF_6, CO, NO_X, SO_X, NMVOCs, BC, OC, NH_3	CO_2, CH_4, N_2O, long-lived F-gases, short-lived F-gases, and SO_X	CO_2, CH_4, N_2O, CO, NO_X, SO_2, NMVOCs, BC, OC, HFC245fa, HFC134a, HFC125, HFC143a, SF_6, C_2F_6, and CF_4
Land Use	Agriculture (crops, livestock, and forests), biomass land use, and land use for wind and/or solar energy	Reduced-form emissions from land-use; no explicit land use sector; assume no net terrestrial emissions of CO_2	Agriculture (crops, pasture, and forests) as well as biomass land use and unmanaged land; the agriculture-land-use module directly determines land-use change emissions and terrestrial carbon stocks.
Population	Exogenous	Exogenous	Exogenous

Table 2.1 Characteristics of the Models, continued

Feature	IGSM (with EPPA Economics Component)	MERGE	MiniCAM
GDP Growth	Exogenous productivity growth assumptions for labor, energy, and land; exogenous labor force growth determined from population growth; endogenous capital growth through savings and investment	Exogenous productivity growth assumptions for labor and energy; exogenous labor force growth determined from population growth; endogenous capital growth through savings and investment	Exogenous productivity growth assumptions for labor; exogenous labor force growth based on population demographics
Energy Efficiency Change	Exogenous	Proportional to the rate of GDP growth in each region	Exogenous
Energy Resources	Oil (including tar sands), shale oil, gas, coal, wind and/or solar, land (biomass), hydro, and nuclear fuel	Conventional oil, unconventional oil (coal-based synthetics, tar sands, and shale oil), gas, coal, wind, solar, biomass, hydro, and nuclear fuel	Conventional oil, unconventional oil (including tar sands and shale oil), gas, coal, wind, solar, biomass (waste and/or residues and crops), hydro, and nuclear fuel (uranium and thorium); includes a full representation of the nuclear fuel cycle
Electricity Technologies	Conventional fossil (coal, gas, and oil), nuclear, hydro, natural gas combined cycle (NGCC) with and without capture, integrated coal gasification with capture, and wind and/or solar, biomass	Conventional fossil (coal, gas, and oil), nuclear, hydro, new coal and gas with and without CCS, other renewables.	Conventional fossil (coal, gas, and oil) with and without capture; integrated gasification combined cycles (IGCCs) with and without capture; NGCC with and without capture; Gen II, III, and IV reactors and associated fuel cycles; hydro, wind, solar, and biomass (traditional and modern commercial)
Conversion Technologies	Oil refining, coal gasification, and bio-liquids	Oil refining, coal gasification and liquefaction, bio-liquids, and electrolysis	Oil refining, natural gas processing, natural gas to liquids conversion, coal, and biomass conversion to synthetic liquids and gases; hydrogen production using liquids, natural gas, coal, biomass; and electrolysis, including direct production from wind and solar, and nuclear thermal conversion
Atmosphere-Ocean	2-dimensional atmosphere with a 3-dimensional ocean general circulation model, resolved at 20 minute time steps, 4° latitude, 4 surface types, and 12 vertical layers in the atmosphere	Parameterized ocean thermal lag	Global multi-box energy balance model with upwelling-diffusion ocean heat transport

Table 2.1 Characteristics of the Models, continued

Feature	IGSM (with EPPA Economics Component)	MERGE	MiniCAM
Carbon Cycle	Biogeochemical models of terrestrial and ocean processes; depends on climate and/or atmospheric conditions with 35 terrestrial ecosystem types	Convolution ocean carbon cycle model assuming a neutral biosphere	Globally balanced carbon-cycle with separate ocean and terrestrial components, with terrestrial response to land-use changes
Natural Emissions	CH_4, N_2O, and weather and/or climate dependent as part of biogeochemical process models	Fixed natural emissions over time	Fixed natural emissions over time
Atmospheric fate of GHGs, pollutants	Process models of atmospheric chemistry resolved for urban and background conditions	Single box models with fixed decay rates. No consideration of reactive gases	Reduced form models for reactive gases and their interactions
Radiation Code	Radiation code accounting for all significant GHGs and aerosols	Reduced form, top-of-the-atmosphere forcing	Reduced form and top-of-the-atmosphere forcing; including indirect forcing effects

SOCIOECONOMIC AND TECHNOLOGY COMPONENTS

Equilibrium, Expectations, and Trade

As can be seen in Table 2.1, the three participating models represent economic activity and associated emissions in a similar way; each divides the world economy into several regions, and further divides each region into economic sectors. In all three, the greatest degree of disaggregation is applied to the various components of energy supply and demand.

The models differ, however, in their representations of the equilibrium structure, the role of future expectations, and in the goods and services traded. MERGE and the EPPA component of IGSM are CGE models, which solve for a consistent set of supply-demand and price equilibria for each good and factor of production that is distinguished in the analysis. In the process, CGE models ensure a balance in each period of income and expenditure and of savings and investment for the economy, and they maintain a balance in international trade in goods and emissions permits. MiniCAM is a partial-equilibrium model, solving for supply-demand and price equilibria within linked energy and agricultural markets. Other economic sectors that

influence the demand for energy and agricultural products and the costs of factors of production in these sectors are represented through exogenous assumptions.

The models also differ in how expectations about the future affect current decisions. The EPPA component of IGSM and MiniCAM are recursive-dynamic, meaning they are solved one period at a time with economic agents modeled as responding to conditions in that period. This behavior is also referred to as myopic because these agents do not consider expected future market conditions in their decisions. The underlying behavioral assumption is that consumers and producers maximize their individual utilities or profits. In MiniCAM, this process is captured through the use of demand and supply functions that evolve over time as a function of evolving economic activity and regional economic development. In IGSM, explicit representative-agent utility and sector production functions ensure that consumer and producer decisions are consistent with welfare and profit maximization. In both of these models, the patterns of emissions mitigation over time in the scenarios that stabilize radiative forcing are imposed through assumptions intended to capture the features of a strategy that, as explained in Section 2.4, would be cost efficient. MERGE, on the other hand, is an inter-temporal optimization model, meaning that all periods are solved simultaneously such that resources and mitigation effort are allocated optimally over time as well as among sectors. Inter-temporal models of this type are often referred to as forward-looking or perfect foresight models because actors in the economy base current decisions not only on current conditions but on future ones, which are assumed to be known with certainty. Simultaneous solution of all periods ensures that agents' expectations about the future are realized in the model solution. MERGE's forward-looking structure allows it to explicitly solve for cost-minimizing emissions pathways, in contrast to MiniCAM and IGSM, which exogenously prescribe emissions mitigation policies over time.

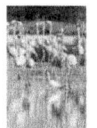

Although all three models also represent international trade in goods and services and include exchange in emissions permits, they differ in the combinations of goods and services traded. In IGSM, all goods and services represented in the model are traded, with electricity trade limited to geographically contiguous regions to the extent that it occurs in the base data. MiniCAM models international trade in oil, coal, natural gas, agricultural goods, and emission permits. MERGE models trade in oil and natural gas, emissions permits, energy-intensive industrial goods, and a single non-energy good representing all other tradable goods and services.

Population and Economic Growth

An increase in the overall scale of economic activity is among the most important drivers of GHG emissions. However, economic growth depends, in part, on growth in population, which in all three models is an exogenously determined input. Although economic activity is an output of the models, its level is largely determined by assumptions about labor productivity and labor force growth, which are also model inputs. Policies to reduce emissions below those in the reference scenarios also affect economic activity, which may be measured as changes in gross domestic product (GDP) or in national consumption. (See Chapter 4, which provides a discussion of the interpretation and limitations of GDP and other welfare measures.)

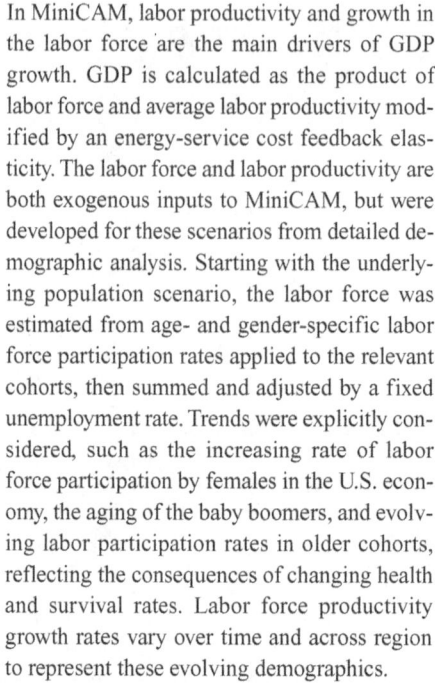

In MiniCAM, labor productivity and growth in the labor force are the main drivers of GDP growth. GDP is calculated as the product of labor force and average labor productivity modified by an energy-service cost feedback elasticity. The labor force and labor productivity are both exogenous inputs to MiniCAM, but were developed for these scenarios from detailed demographic analysis. Starting with the underlying population scenario, the labor force was estimated from age- and gender-specific labor force participation rates applied to the relevant cohorts, then summed and adjusted by a fixed unemployment rate. Trends were explicitly considered, such as the increasing rate of labor force participation by females in the U.S. economy, the aging of the baby boomers, and evolving labor participation rates in older cohorts, reflecting the consequences of changing health and survival rates. Labor force productivity growth rates vary over time and across region to represent these evolving demographics.

In MERGE and the EPPA component of IGSM, the labor force and its productivity, while extremely important, are not the only factors determining GDP. Savings and investment and productivity growth in other factors (e.g., materials, land, labor, and energy) variously contribute as well. IGSM and MERGE use population directly as a measure of the labor force and apply assumptions about labor productivity change that are appropriate for that definition.

Energy Demand

In all three models, energy demands are represented regionally and driven by regional economic activity. As a region's economic activity increases, its corresponding demand for energy services rises. Energy demand is also affected by assumptions about changes in technology, in the structure of the economy, and in other economic conditions (see Section 2.2.5). Similarly, all the models represent the way demand will respond to changes in price. The formulation of price response is particularly important in the construction of stabilization scenarios because the imposition of a constraint on carbon emissions will require the use of more expensive energy sources with lower emissions and will, therefore, raise the consumer price of all forms of energy.

The demand for energy is derived from demands for other goods and services in all three IAMs. However, the models differ in the way they derive their energy demands. In IGSM each good- or service-producing sector demands energy. The production sector is an input-output structure in which every industry (including the energy sector) supplies its outputs as inputs to intermediate production in other industries and for final consumption. Households have separate demands for automobile fuel and for all other energy services. Each final demand sector can use electricity, liquid fuels (petroleum products or biomass liquids), gas, and coal; fuel for automobiles is limited to liquids. MiniCAM is similar in that each MiniCAM sector demands energy. Energy is demanded by both final consumers and transforming sectors. In MiniCAM, there are three final energy consumption sectors – buildings, industry, and transport – which consume electricity and energy products such as coal, biomass, refined liquid fuels, methane, and hydrogen. In addition, energy is demanded by energy-producing and refining sectors, power generators, and hydrogen producers, whose demands in turn are derived from the demands arising in the final energy consumption sectors. MERGE is similar to IGSM except that its interindustry transactions are aggregated into a single, non-energy-production sector for each region from which demands for fuels (oil, gas, coal, and bioenergy) and electricity are derived. The power generation sector's demands for energy are derived from the economy's demand for electricity.

Energy Resources

The future availability of energy resources, particularly of exhaustible fossil fuels, is an important determinant of energy use and emissions, so all three of the participating models provide explicit treatments of the underlying resource base. All three include empirically based estimates of in-ground resources of oil, coal, and natural gas that might ultimately be available, along with a model of the costs of extraction. The levels of detail in the different models are shown in Table 2.1. Each of the models includes both conventional and unconventional sources in its resource base and represents the process of exhaustion of resources

by an increasing cost of exploitation. That is, lower-cost resources are utilized first so that the costs of extraction rise as the resources are depleted. The models differ, however, in the way they represent the increasing costs of extraction. MiniCAM divides the resource base for each fossil fuel into discrete grades with increasing costs of extraction, along with an exogenous technological change parameter that lowers extraction costs over time. MERGE has similar differential grades for oil and gas, but assumes that the coal base is more than sufficient to meet potential demand and that exogenous technological improvements in extraction will be minimal. For these reasons, MERGE represents coal as having a constant cost over time irrespective of utilization. IGSM models resource grades with a continuous function, separately identifying conventional oil, shale oil, natural gas, and coal. Fuel-producing sectors are subject to economy-wide technical progress (e.g., increased labor productivity growth), which partly offsets the rise in extraction costs. The models all incorporate tar sands and unconventional gas (e.g., tight gas and coal-seam gas) in the grade structure for oil and natural gas, and each also includes the potential development of shale oil.

The models seek to represent all resources that could be available as technology and economic conditions vary over time and across simulations. Thus, they represent conditions under which currently unused resources could be economically exploited due to advances in technology or higher prices driven by increasing demands. Generally, then, the modeling groups define a resource base that is more expansive than, for example, that of the U.S. Geological Survey, which estimates technological and economic feasibility only at current technology and prices. However, differences exist in the treatments of potentially available resources. MiniCAM includes a detailed representation of the nuclear power sector, including uranium and thorium resources; nuclear fuel fabrication; reactor technology options; and associated fuel-cycle cycles, including waste, storage, and fuel reprocessing. IGSM and MERGE assume that the uranium resources used for nuclear power generation are sufficient to meet likely use and, therefore, do not explicitly model their depletion.

The treatment of wind and solar resources also differs among the models. IGSM represents the penalty for intermittent supply by modeling wind and solar as imperfect substitutes for central station generation, where the elasticity of substitution implies a rising cost as these resources supply a larger share of electricity supply. Land is also an input, and the regional cost of wind and solar energy is based on estimates of regional resource availability and quality. MERGE represents these resources as having a fixed cost, but it applies upper limits on the proportion of these resources in the electricity system, representing limits on the integration of these resources into the grid. MiniCAM represents wind and solar technologies as extracting power from graded, regional, renewable resource bases. Variation in resource availability across diurnal and annual cycles affects market penetration of these technologies. As wind and solar technologies achieve larger fractions of the total power generation system, storage and ancillary power production capacity are required, which in turn affects the cost of power generation and technology choice.

IGSM and MiniCAM model biomass production as competing for agricultural land. Increasing production leads to increasing land rent, representing the scarcity of agricultural land, and thus, to increasing cost of biomass as production expands. MiniCAM also has a separate set of regional supply functions for biomass supplied from waste and residue sources. In these scenarios, MERGE represents biomass as a graded resource. Two grades of biomass are included, with fixed costs for each. The total supply from the first, less-expensive grade is limited, but the second, more-expensive grade is allowed to compete unhindered in the market.

Technology and Technological Change

Technology is the broad set of processes covering know-how, experience, and equipment used by humans to produce services and transform resources. In the three models participating in this scenario, the relationship between things that are produced and things that are used in the production process are represented mathematically. In the jargon of the models, the relationship between things that are produced and things that are used in the production process is

referred to as a production function.

The three modeling groups differed substantially in their representation of technology depending on their overall design objectives. Differences also resulted from data limitations and computational feasibility, which force trade-offs between the inclusion of engineering detail and the representation of the interaction among the segments of a modern economy that determines supply, demand, and prices (see Box 2.1).

All three of the models applied here follow a hybrid approach to the representation of energy technology, involving substantial detail in some areas and more aggregate representations in others, and some of the choices that flow from the distinct design of each can be seen in Table 2.1. They represent energy demand, as described in Section 2.2.3, with the application of an autonomous energy efficiency improvement (AEEI) factor to represent non-price-induced trends in energy use. However, AEEI parameter values are not directly comparable across the models because each has a unique representation of the processes that together explain the multiple forces that have contributed historically to changes in the energy intensity of economic activity. In IGSM and MERGE, the AEEI captures non-price changes (including structural change not accounted for in the models) that can be energy using rather than energy saving. MERGE represents the AEEI as a function of GDP growth in each region. MiniCAM captures shifts among fuels through differing income elasticities, which change over time, and separately represents AEEI efficiency gains.

Other areas shown in Table 2.1 where there are significant differences among the models are in energy conversion – from fossil fuels or renewable sources to electricity and from solid fossil fuels or biomass to liquid fuels or gas. In IGSM, discrete energy technologies are represented as energy supply sectors contained within the input-output structure of the economy. Those sources of fuels and electricity that now dominate supply are represented as production functions with the same basic structure as the other sectors of the economy. Technologies that may play a large role in the future (e.g., power plants with CCS or oil from shale) are introduced as discrete technologies using a production function structure similar to that for existing pro-

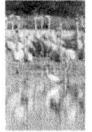

> **BOX 2.1 Top-Down, Bottom-Up, and Hybrid Modeling**
>
> The models used in energy and environmental assessments are sometimes classified as either top-down or bottom-up in structure, a distinction that refers to the way they represent technological options. A top-down model uses an aggregate representation of how producers and consumers can substitute non-energy inputs for energy inputs or relatively energy-intensive goods for less energy-intensive goods. Often, these tradeoffs are represented by aggregate production functions or by utility functions that describe consumers' willingness and technical ability to substitute among goods.
>
> The bottom-up approach begins with explicit technological options, and fuel substitution or changes in efficiency occur as a result of discrete changes from one specific technology to another. The bottom-up approach has the advantage of being able to represent explicitly the combination of outputs, inputs, and emissions of types of capital equipment used to provide consumer services (e.g., a vehicle model or building design) or to perform a particular step in energy supply (e.g., a coal-fired powerplant or wind turbine). However, a limited number of technologies are often included, which may not well represent the full set of possible options that exist in practice. Also, in a pure bottom-up approach, the demands for particular energy services are often characterized as fixed (unresponsive to price), and the prices of inputs such as capital, labor, energy, and materials are exogenous.
>
> On the other hand, the top-down approach explicitly models demand responsiveness and input prices, which usually require the use of continuous functions to model at least some parts of the available technology set. The disadvantage of the latter approach is that production functions of this form will poorly represent switch points from one technology to another – as from one form of electric generation to another or from gasoline to biomass blends as vehicle fuel. In practice, the vast majority of models in use today, including those applied in this scenario, are hybrids in that they include substantial technological detail in some sectors and more aggregate representations in others.

duction sectors and technologies. They are subject to economy-wide productivity improvements (e.g., labor, land, and energy productivity), with the effect on cost depending on the share of each factor in the technology production function. MERGE and MiniCAM also characterize energy-supply technologies in terms of discrete technologies. In the MERGE scenarios in this research, technological improvements are captured by allowing for the introduction of more advanced technologies in future periods. In the MiniCAM scenarios, the cost and performance of technologies are assumed to improve over time, and new technologies become available in the future. Similar differences among the models hold for other conversion technologies, such as coal gasification, coal liquefaction, or liquids from biomass.

The entry into the market of new sources and their levels of production by region are determined endogenously in all three models and depend on the relative costs of supply. It should be emphasized that the versions of the models used in this research do not explicitly represent the processes of technological change, for example, public and private R&D, spillovers from innovation in other economic sectors, and learning-by-doing. A number of recent efforts have been made to incorporate such processes and their effects as an endogenous component of modeling exercises. In most cases, these studies have not been applied to models of the complexity needed to meet the requirements of this scenario product.

Because of the differences in structure among these models, there is no simple technology-by-technology comparison of performance and cost across particular sources of supply or technological options. This situation exists for a variety of reasons. First, cost is an output of the three models and not an input. In the three models here technologies are defined in many cases not in terms of some exogenously specified cost, but rather as a function of inputs whose prices change across simulations and over time. The three models differ in many regards. Each model defines the scope of a technology differently. Sectoral definitions, technology definitions, and data sources all vary across the three models. For example, one model has a service sector while another has a buildings sector. There is then, no common definition for technologies, technology descriptors and hence for a set of comparable costs. The detailed scenario documentation for each of the three modeling groups provides more information about the technology assumptions employed by three modeling groups. These are documented in Paltsev et al. (2005) for IGSM and in Clarke et al. (2007) for MiniCAM. Assumptions for MERGE are included in the version of the model posted at http://www.stanford.edu/group/MERGE.

The influence of differing technology specifications and assumptions is evident in the scenarios discussed in Chapters 3 and 4. For example, in the absence of efforts to control GHG emissions, motor fuel is drawn ever more heavily from high-emitting sources. Oil from shale comes in under the resource and technology assumptions used in the IGSM scenarios, whereas liquids from coal figure prominently in the MERGE scenarios, and the MiniCAM scenarios include an intermediate mix of both. Furthermore, because each model assumes market mechanisms operate efficiently, the marginal cost of reducing GHG emissions – that is the cost of reducing the last tonne of GHG – is equal to the price of carbon in every technology employed in every sector and in every country of the world. When stabilization conditions are imposed, CCS takes on a key role in all the scenarios over the time period considered in this research. Nuclear power contributes heavily in MERGE and in MiniCAM scenarios, whereas the potential role of this technology is overridden in the IGSM scenarios by an assumption of non-climate restraints on expansion due to concerns over issues such as safety, waste, and proliferation. Finally, although differences in emissions in the reference scenario contribute to variations in the difficulty of achieving stabilization, alternative assumptions about technological improvements also play a prominent role.

Land Use and Land-Use Change

The models used in this research were developed originally with a focus on energy and fossil carbon emissions. The integration of the terrestrial biosphere, including human activity,

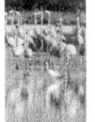

into the climate system is less highly developed. Each model represents the global carbon cycle, including exchanges among the atmosphere, natural vegetation, and soils; the effects of human land use and responses to carbon policy; and feedbacks to the global climate. No model represents all of these possible responses and interactions, and the level of detail varies substantially among the models. For example, the models differ in their handling of natural vegetation and soils and in their responses to change CO_2 concentrations and climate. Furthermore, land-use practices (e.g., low- or no-till agriculture and biomass production) and changes in land use (e.g., afforestation, reforestation, or deforestation) that influence GHG emissions and the sequestration of carbon in terrestrial systems are handled at different levels of detail. Indeed, improved two-way linking of global economic and climate analysis with models of physical land use (land use responding to climate and economic pressures and climate responding to changes in the terrestrial biosphere) is the subject of ongoing research in these modeling groups.

In IGSM, land is an input to agriculture, biomass production, and wind and/or solar energy production. Agriculture is a single sector that aggregates crops, livestock, and forestry. Biomass energy production is modeled as a separate sector, which competes with agriculture for land. Markets for agricultural goods and biomass energy are international, and demand for these products determines the price of land in each region and its allocation among uses. In other sectors, returns to capital include returns to land, but the land component is not explicitly identified. Anthropogenic emissions of GHGs (importantly, CH_4 and N_2O) are estimated within IGSM as functions of agricultural activity and assumed levels of deforestation. The response of terrestrial vegetation and soils to climate change and CO_2 increase is captured in the Earth system component of the model, which provides a detailed treatment of biogeochemical and land-surface properties of terrestrial systems. However, the biogeography of natural ecosystems and human uses remains unchanged over the simulation period, with the area of cropland fixed to the pattern of the early 1990s. Balance in the carbon cycle between ocean uptake, land-use and land-use change, and anthropogenic emissions is achieved in

IGSM with an adjustment factor to ensure that the recent trend in atmospheric CO_2 increase is replicated. This adjustment factor is best interpreted as what carbon uptake due to forest regrowth must have been, given the representation of terrestrial and ocean systems in IGSM. The need for such an adjustment factor reflects the continuing scientific uncertainty in the carbon cycle. In other words, with fossil emissions and concentrations relatively well known, the total uptake is known but the partitioning of the uptake between terrestrial and ocean systems is uncertain (Sabine et al. 2004). IGSM does not simulate carbon price-induced changes in carbon sequestration (e.g., reforestation and tillage), and change among land-use types in the EPPA component of IGSM is not fed to the terrestrial biosphere component of the model.

The MERGE modeling group assumed a neutral terrestrial biosphere across all scenarios. That is, it is assumed that the net CO_2 exchange with the atmosphere by natural ecosystems and managed systems – the latter including agriculture, deforestation, afforestation, reforestation, and other land-use change – sums to zero.

MiniCAM includes a model that allocates the land area in a region among various components of human use and unmanaged land – with changes in allocation over time in relation to income, technology, and prices – and estimates the CO_2 emissions (or sinks) that result. Land conditions and associated emissions are parameterized for a set of regional sub-aggregates. The supply of primary agricultural production (four food crop types, pasture, wood, and commercial biomass) is simulated regionally with competition for a finite land resource based on the average profit rate for each good potentially produced in a region. In stabilization scenarios, the value of carbon stored in the land is added to this profit, based on the average carbon content of different land uses in each region. This allows carbon mitigation policies to explicitly extend into land and agricultural markets. The model is solved by clearing a global market for primary agricultural goods and regional markets for pasture. The biomass market is cleared with demand for biomass from the energy component of the model. Exogenous assumptions are made for the rate of intrinsic increase in agricultural productivity, although net productivity

can decrease in the case of expansion of agricultural lands into less productive areas (Sands and Leimbach 2003). Unmanaged land can be converted to agro-forestry, which in general leads to net CO_2 emissions from tropical regions in the early decades. Emissions of non-CO_2 GHGs are tied to relevant drivers, for example, with CH_4 from ruminant animals related to beef production. MiniCAM thus treats the effects on carbon emissions of gross changes in land use (e.g., from forests to biomass production) using an average emission factor for such conversion. The pricing of carbon stocks in the model provides a counterbalance to increasing demand for biomass crops in stabilization scenarios.

Emissions of CO_2 and Non-CO_2 Greenhouse Gases

In all three models, the main source of CO_2 emissions is fossil fuel combustion, which is computed on the basis of the carbon content of each of the underlying resources: oil, natural gas, and coal. Special adjustments are made to account for emissions associated with the additional processing required to convert coal, tar sands, and shale sources into products equivalent to those from conventional oil. Other industrial CO_2 emissions also are included, primarily from cement production.

As required for this research, all three models include representations of emissions and abatement of CH_4, N_2O, HFCs, PFCs, and SF_6 (plus aerosols and other substances not considered in this scenario). The models use somewhat different approaches to represent abatement of non-CO_2 GHGs. IGSM includes the emissions and abatement possibilities directly in the production functions of the sectors that are responsible for emissions of the different gases. Abatement possibilities are represented by substitution elasticities in a nested structure that encompasses GHG emissions and other inputs, benchmarked to reflect bottom-up studies of abatement potential. This construction is parallel to the representation of fossil fuels in production functions, where abatement potential is similarly represented by the substitution elasticity between fossil fuels and other inputs, with the specific set of substitutions governed by the nest structure. Abatement opportunities vary by sector and region.

In MERGE, CH_4 emissions from natural gas use are tied directly to the level of natural gas consumption, with the emissions rate decreasing over time to represent reduced leakage during the transportation process. Non-energy sources of CH_4, N_2O, HFCs, PFCs, and SF_6 are based largely on the guidelines provided by the EMF Study No. 21 on Multi-Gas Mitigation and Climate Change (de la Chesnaye and Weyant 2006). The EMF developed baseline projections from 2000 through 2020. For all gases but N_2O and CO_2, the baseline for beyond 2020 was derived by extrapolation of these estimates. Abatement cost functions – the relationship between levels of emissions reductions and the costs of these reductions – for these two gases are also based on EMF 21, which provided estimates of the abatement potential for each gas in each of 11 cost categories in 2010. These abatement cost curves are directly incorporated in the model and extrapolated after 2010 following the baseline. There is also an allowance for technical advances in abatement over time.

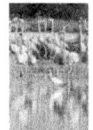

MiniCAM calculates emissions of CH_4, N_2O, and seven categories of industrial sources for HFCs, PFCs, and SF_6. Emissions are determined for over 30 sectors, including fossil fuel production, transformation, and combustion; industrial processes; land use and land-use change; and urban emissions. For details, see Smith (2005) and Smith and Wigley (2006). Emissions are proportional to driving factors appropriate for each sector, with emissions factors in many sectors decreasing over time according to an income-driven logistic formulation. Marginal abatement cost (MAC) curves from the EMF-21 study are applied, including shifts in the curves for CH_4 due to changes in natural gas prices. Any below-zero reductions in MAC curves are assumed to apply in the reference scenario.

EARTH SYSTEMS COMPONENTS

The Earth system components of the models represent the response of the atmosphere, ocean, and terrestrial biosphere to emissions and increasing concentrations of GHGs and other substances. Representation of these processes, including the carbon cycle (Box 2.2), is necessary to determine emissions paths consistent with stabilization because these systems

determine how long each of these substances remains in the atmosphere and how they interact in altering the Earth's radiation balance. Each model includes such physical-chemical-biological components, but incorporates different levels of detail. The most elaborated Earth system components are found in IGSM (Sokolov et al. 2005), which falls in a class of models referred to as Earth System Models of Intermediate Complexity (Claussen et al. 2002). These are models that fall between the full three-dimensional atmosphere-ocean general circulation models (AOGCMs) and energy balance models with a box model of the carbon cycle. The Earth system components of MERGE and MiniCAM fall in the class of energy balance-carbon cycle box models. Table 2.1 shows how each of the models treat different components of the Earth systems.

IGSM has explicit spatial detail, resolving the atmosphere into multiple layers and by latitude, and it includes a terrestrial vegetation model with multiple vegetation types that are also spatially resolved. A version of IGSM with a full three-dimensional ocean model was used for this scenario, and it includes temperature-dependent uptake of carbon. IGSM models atmospheric chemistry, resolved separately for urban (i.e., heavily polluted) and background conditions. Processes that move carbon into or out of the ocean and vegetation are modeled explicitly. IGSM also models natural emissions of CH_4 and N_2O, which are weather and/or climate-dependent. The model includes a radiation code that computes the net effect of atmospheric concentrations of the GHGs studied in this research. Also included in the global forcing is the effect of changing ozone and aerosol levels, which result from emissions of CH_4 and non-GHGs, such as NO_X and volatile organic hydrocarbons; SO_X; black carbon; and organic carbon from energy, industrial, agricultural, and natural sources.

The carbon cycle in MERGE relates emissions to concentrations using a convolution ocean carbon-cycle model and assuming a neutral biosphere (i.e., no net CO_2 exchange). It is a reduced-form carbon cycle model developed by Maier-Reimer and Hasselmann (1987). Carbon emissions are divided into five classes, each with different atmospheric lifetimes. The be-

---

BOX 2.2 The Carbon Cycle

Although an approximate atmospheric lifetime is sometimes calculated for CO_2, the term is potentially misleading because it implies that CO_2 put into the atmosphere by human activity always declines over time by some stable removal process. In fact, the calculated concentration of CO_2 is not related to any mechanism of destruction, or even to the length of time an individual molecule spends in the atmosphere, because CO_2 is constantly exchanged between the atmosphere and the surface layer of the ocean and with vegetation. Instead, it is more appropriate to think about how the quantity of carbon that the Earth contains is partitioned between stocks of in-ground fossil resources, the atmosphere (mainly as CO_2), surface vegetation and soils, and the surface and deep layers of the ocean. When stored carbon is released into the atmosphere, either from fossil or terrestrial sources, atmospheric concentrations of CO_2 increase, leading to disequilibrium with the ocean, and more carbon is taken up than is cycled back. For land processes, vegetation growth may be enhanced by increases in atmospheric CO_2, and this change could augment the stock of carbon in vegetation and soils. As a result of the ocean and terrestrial uptake, only about half of the carbon currently emitted remains in the atmosphere. Over millennial time scales, oceans would continue to remove carbon until a large fraction, presently about 80%, would ultimately be removed to the oceans, leaving about 20% as a permanent increase in the atmospheric CO_2 concentration. But this large removal only occurs because current levels of emissions lead to substantial disequilibrium between atmosphere and ocean. Lower emissions would lead to less uptake, as atmospheric concentrations come into balance with the ocean and interact with the terrestrial system. Rising temperatures themselves will reduce uptake by the ocean, and will affect terrestrial vegetation uptake, processes that the models in this scenario variously represent.

An important policy implication of these carbon-cycle processes as they affect stabilization scenarios is that stabilization of emissions near the present level will not lead to stabilization of atmospheric concentrations. CO_2 concentrations were increasing in the 1990s at just over 3 ppmv per year, an annual increase of 0.8%. Thus, even if societies were able to stabilize emissions at current levels, atmospheric concentrations of CO_2 would continue to rise. As long as emissions exceed the rate of uptake, even very stringent abatement will only slow the rate of increase.

havior of the model compares favorably with atmospheric concentrations provided in the IPCC's Third Assessment Report (TAR) (IPCC 2001) when the same SRES scenarios of emissions are simulated in the model (Nakicenovic et al. 2000). MERGE models the radiative effects of GHGs using relationships consistent with summaries by the IPCC, and applies the median aerosol forcing from Wigley and Raper (2001). The aggregate effect is obtained by summing the radiative forcing effect of each gas.

MERGE's physical Earth system component is embedded in the inter-temporal optimization framework, thus allowing solution of an optimal allocation of resources through time, accounting for damages related to climate change, or optimizing the allocation of resources with regard to other constraints such as concentrations, temperature, or radiative forcing. In this research, the second of these capabilities is applied, with a constraint on radiative forcing (see

Chapter 4). In contrast, the IGSM and MiniCAM Earth system models are driven by emissions as simulated by the economic components. In that regard, they are simulations rather than optimization models.

MiniCAM uses the MAGICC model (Wigley and Raper 2001, 2002) as its biophysical component. MAGICC is an energy-balance climate model that simulates the energy inputs and outputs of key components of the climate system (sun, atmosphere, land surface, and ocean) with parameterizations of dynamic processes such as ocean circulations. It operates by taking anthropogenic emissions from the other MiniCAM components, converting these to global average concentrations (for gaseous emissions), then determining anthropogenic radiative forcing relative to preindustrial conditions, and finally computing global mean temperature changes. The carbon cycle is modeled with both terrestrial and ocean components. The terrestrial component includes CO_2 fertilization and temperature feedbacks; the ocean component is a modified version of the Maier-Reimer and Hasselmann (1987) model that also includes temperature effects on the terrestrial biosphere. Net land-use change emissions from the MiniCAM's land-use change component are fed into MAGICC so that the global carbon cycle is consistent with the amount of natural vegetation. Reactive gases and their interactions are modeled on a global-mean basis using equations derived from results of global atmospheric chemistry models (Wigley et al. 2002).

In MiniCAM, global mean radiative forcing for CO_2, CH_4, and N_2O are determined from GHG concentrations using analytic approximations. Radiative forcing for other GHGs are taken to be proportional to concentrations. Radiative forcing for aerosols (for sulfur dioxide and for black and organic carbon) are taken to be proportional to emissions. Indirect forcing effects, such as the effect of CH_4 on stratospheric water vapor, are also included. Given radiative forcing, global mean temperature changes are determined by a multiple box model with an upwelling-diffusion ocean component. The climate sensitivity is specified as an exogenous parameter. MAGICC's ability to reproduce the global mean temperature change results of AOGCMs has been demonstrated (Cubasch et al. 2001, Raper and Gregory 2001).

Although aerosols and ozone are not included in the computation of the radiative forcing targets that are the focus of these scenarios, they are nonetheless included in these scenarios as noted above. That is, the radiative forcing stabilization levels identified in Table 1.2 and the radiative forcing levels reported in subsequent chapters account for only that part of radiative forcing due to those GHGs covered by the target. The models can simulate total radiative forcing including additional positive forcing from ozone and dark aerosols and negative forcing from sulfate aerosols. As shown by Prinn et al. (In Press), even for very large changes in emissions related to these substances, the temperature effect is small, in large part because aerosols and ozone have offsetting cooling and warming effects. To the extent temperature is affected by these substances, however, they have a small, indirect influence on the scenarios be-

cause trace gas cycles are climate-dependent. For example, climate affects vegetation and ocean temperature and, thus, carbon uptake, and natural emissions of CH_4 and N_2O, and the lifetime of CH_4 also depends on climate. Because the net effect of these substances on temperature is small, the feedback effect on trace gas cycles also is very small. However, to the extent these feedbacks are represented in the models as discussed above, they are included in the calculation of required emissions reduction because the temperature paths, while not reported here, are simulated in the models and affect the CO_2 and non-CO_2 GHG concentrations. By the same token, the gases included under the Montreal Protocol, which are being phased out, are nonetheless included in these models and exert some influence on temperature.

Note that although the models used in this research have capabilities to evaluate various climate change effects, with few exceptions, they do not include the consequences of such feedback effects as: temperature on home heating and cooling requirements; local climate change on agricultural productivity; CO_2 fertilization on agricultural productivity (though a CO_2 fertilization effect is included in the terrestrial carbon cycle models employed by IGSM and MiniCAM); climate on water availability for applications ranging from crop growing to power plant cooling. Such improvements are left to future research.

Reference
Scenarios

In the reference scenarios, energy consumption grows significantly and the energy system continues to rely on fossil fuels, leading to an increase in CO₂ emissions of roughly 3 to 3½ times the present level by 2100. Combined with increases in the non-CO₂ GHGs and net uptake by the ocean and terrestrial biosphere, radiative forcing from the GHGs considered in this research reaches 6.4 W/m² to 8.6 W/m² from preindustrial by 2100.

INTRODUCTION

This chapter introduces the reference scenarios developed by the three modeling groups. These scenarios are plausible future paths, not predictions, for by the very nature of their construction they lack the features of predictions or best-judgment forecasts. For example, they assume that in the post-2012 period existing measures to address climate change expire and are never renewed or replaced, which is an unlikely occurrence. Rather, they have been developed as points of departure to highlight the implications for energy use and other human activities of the stabilization of radiative forcing. Each of the modeling groups could have created a range of other plausible reference scenarios by varying assumptions about rates of economic growth, the cost and availability of alternative energy options, assumptions about non-climate environmental regulations, and so forth.

Other than to standardize reporting conventions and GHG emissions mitigation policies (or lack thereof), the three modeling groups developed their reference scenarios independently as each judged appropriate. As noted in Chapter 2, the three models were developed with somewhat different original design objectives. They differ in (a) their inclusiveness, (b) their specifications of key aspects of economic structure, and (c) their choice of values for key parameters. These choices then lead to different characterizations of the underlying economic and physical systems that these models represent.

Moreover, even if the models were identical in structure, the independent choice of key assumptions by the modeling groups leads to differences among scenarios. For example, as will be discussed, the reference scenarios differ in their specification of the technical details of virtually every aspect of the future global energy system, ranging from the cost and availability of oil and natural gas to the prospects for nuclear power. These differences affect emissions in the reference scenarios and the nature and cost of stabilization regimes.

Finally, the modeling groups did not attempt to harmonize assumptions about non-climate-related policies. Such differences matter both in the reference and stabilization scenarios. For example, the MiniCAM reference scenario assumes a larger effect of CH_4 emission-control technologies deployed for economic reasons, which leads to lower reference scenario CH_4 emissions than in the reference scenarios from the other modeling groups. Similarly, the IGSM modeling group assumed that non-climate concerns would limit the deployment of nuclear power, while the MERGE and MiniCAM modeling groups assumed that nuclear power would be allowed to participate in energy markets on the basis of energy cost alone.

This variation in modeling approaches and assumptions is one of the strengths of this research, for the resulting differences across scenarios can help shed light on the implications of differing assumptions about the way key forces may evolve over time. It also provides three independent starting points for consideration of stabilization goals.

Although there are many reasons to expect that the three reference scenarios would be different, it is worth noting that the modeling groups met periodically during the research process to review progress and to exchange information. Thus, while not adhering to any formal protocol of standardization, the three reference scenarios are not entirely independent either.

Development of a reference scenario involves the elaboration of one path from among a range of uncertain outcomes. Thus, it should be further emphasized that the three reference scenarios were not designed in an attempt to span the full range of potential future conditions or to shed light on the probability of the occurrence of future events. That is a much more ambitious undertaking than the one reported here.

The remainder of this chapter describes the reference scenarios developed by the three modeling groups working forward from underlying

drivers to implications for radiative forcing. (Chapter 4 proceeds in the other direction, imposing the stabilization levels on radiative forcing and exploring the implications.) The presentation begins with a summary of the underlying socioeconomic assumptions, most notably for population and economic growth. There follows a discussion of the evolution of the global energy system over the twenty-first century in the absence of additional GHG controls and discusses the associated prices of fuels. The energy sector is the largest but not the only source of anthropogenic GHG emissions. Also important is the net uptake or release of CO_2 by the oceans and the terrestrial biosphere. The next section shows how the three reference scenarios handle this aspect of the interaction of human activity with natural Earth systems. Finally, the anthropogenic emissions are described, taking into account both the energy sector and other sources, such as agriculture and various industrial activities. This last section draws together all these various components to present reference scenarios of the consequences of anthropogenic emissions and the processes of CO_2 uptake and non-CO_2 gas destruction for the ultimate focus of the research: atmospheric concentrations and global radiative forcing.

SOCIOECONOMIC ASSUMPTIONS

GHGs are a product of modern life. Population increase and economic activity are major determinants of the scale of human activities and ultimately of anthropogenic GHG emissions. In the reference scenarios, the global population rises from 6 billion in the year 2000 to between 8.6 and 9.9 billion in 2100. Economic activity grows through 2100 across the globe. Developed nations continue to expand their economies at historical rates, and developing nations make significant progress toward improved standards of living.

Reference scenarios are grounded in a larger demographic and economic story. Each uses population as the basis for developing scenarios of the scale and composition of economic activity for each region. For population assumptions, the IGSM modeling group adopted a regionally detailed United Nations (U.N.) projection for the period 2000-2050 (UN 2001) and extended this scenario to 2100 using information from a longer-term U.N. study (UN 2000). The Mini-CAM assumptions are based on a median scenario by the U.N. (UN 2005) and a Millennium Assessment Techno-Garden Scenario from the International Institute for Applied Systems Analysis (O'Neill 2005). Near-term population assumptions for the MERGE scenarios come from the Energy Information Administration's International Energy Outlook.

Population increases substantially across the scenarios by the end of the century, but all of the scenarios portray the population growth rate as slowing to near zero, if not turning negative, by the end of the century (Table 3.1 and Figure 3.1). As a result, by 2050 more than 75% of all the change between the year 2000 and 2100 has occurred. A demographic transition from high birth and death rates to low death rates and eventually to low birth rates is a feature of most demographic scenarios, reflecting assumptions that birth rates will decline to replacement levels or below. For some countries, birth rates are

IGSM Population by Region (million)

Region	2000	2020	2040	2060	2080	2100
U.S.	283	334	379	396	395	393
Western Europe	390	388	368	331	302	289
Japan	127	126	116	113	118	119
Former Soviet Union	291	278	260	243	234	230
Eastern Europe	97	91	83	74	67	64
China	1282	1454	1500	1429	1365	1334
India	1009	1291	1503	1610	1635	1643
Africa	793	1230	1749	2163	2390	2500
Latin America	419	538	627	678	701	713
Rest of the World	1366	1848	2269	2521	2614	2652

MERGE Population by Region (million)

Region	2000	2020	2040	2060	2080	2100
U.S.	276	335	335	335	335	335
Western Europe	390	397	397	397	397	397
Japan	127	126	126	126	126	126
Former Soviet Union/Eastern Europe	411	393	393	393	393	393
China	1275	1429	1478	1493	1498	1499
India	1017	1312	1427	1472	1489	1496
Africa/Latin America/Rest of World	2566	3538	4209	4677	5003	5228

MiniCAM Population by Region (million)

Region	2000	2020	2040	2060	2080	2100
U.S.	283	334	371	396	412	426
Western Europe	457	486	481	456	421	399
Japan	127	127	121	113	103	95
Former Soviet Union	283	284	283	275	261	253
Eastern Europe	124	119	111	100	87	80
China	1385	1578	1591	1506	1407	1293
India	1010	1312	1472	1513	1443	1300
Africa	802	1197	1521	1763	1893	1881
Latin America	525	670	786	869	929	952
Rest of World	1055	1454	1779	1976	2012	1918

**Table 3.1.
Population (million) by Region Across Models, 2000-2100.**

Regional aggregations are different in the three models. For example, MiniCAM includes Turkey in Western Europe, but IGSM and MERGE do not.

already below replacement levels, and just maintaining these levels would result in population decline for these countries. A key uncertainty in all demographic scenarios is whether a transition to less-than-replacement levels is a more or less permanent feature of those countries where it has occurred and whether such a pattern will be repeated in other countries.

The differences among the scenarios lie in nuances of this pattern. The MiniCAM scenarios exhibit a peak in global population around the year 2070 at slightly more than 9 billion people, after which the population declines to 8.6 billion. The MERGE and IGSM scenarios, on the other hand, both employ demographic assumptions by which the global population stabilizes but does not decline during this century. By 2100, populations range from 8.6 to 9.9 billion across the scenarios, which is an increase of roughly 40% to 65% from the 6 billion on Earth in 2000. In total, the difference between the demographic scenarios is relatively small: they differ by only 3% in 2030 and by less than 10% until after 2080.

The variation in population among the scenarios is greater for the U.S. than for the globe. The U.S. population (Figure 3.1) increases from about 280 million in the year 2000 to between 335 million and 425 million by 2100. Although the MiniCAM global population is the lowest of

the three scenarios in 2100, it is the highest for the U.S. The higher U.S. population in MiniCAM reference scenarios compared to the scenarios from the other two modeling groups can be traced to different assumptions about net migration.

As discussed in Chapter 2, GDP, while ostensibly an output of all three models, is in fact largely determined by assumptions about labor productivity and labor force growth, which are model inputs. None of the three modeling groups began with a GDP goal and derived sets of input factors that would generate that level of activity. Rather, each began with assessments about potential growth rates in labor productivity and labor force and used these, through differing mechanisms, to compute GDP. In MiniCAM, labor productivity and labor force growth are the main drivers of GDP growth. In MERGE and IGSM, savings and investment and productivity growth in other factors (e.g., materials, land, and energy) contribute as well. All three models derive labor force growth from the underlying assumptions about population.

The alternative scenarios of population and productivity growth lead to differences among the three reference scenarios in U.S. GDP growth (Figure 3.2). There is relatively little difference among the three trajectories through the year 2020. After 2020, however, the scenarios diverge, with the lowest scenario of U.S. GDP

Figure 3.1. World and U.S. Population Across Models. Assumed growth in global and U.S. population is similar among the three models. Global population in 2100 spans a range from about 8.5 to 10 billion. U.S. population in 2100 spans a range from less than .35 to over .45 billion.

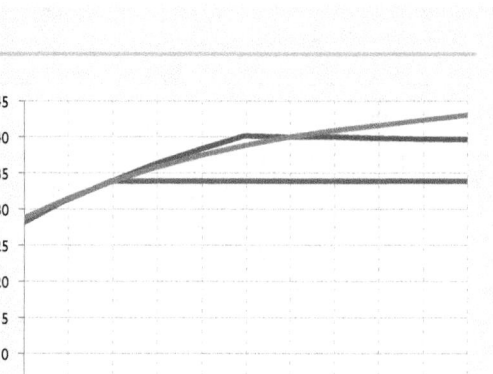

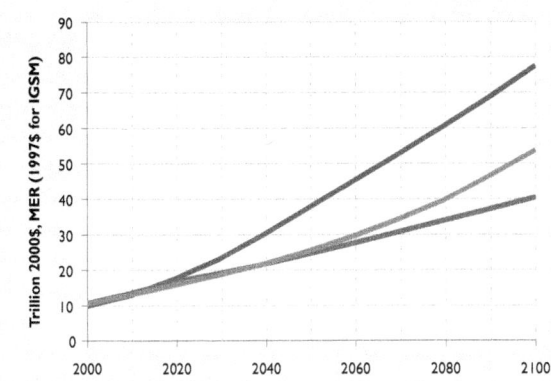

Figure 3.2. U.S. GDP Across Reference Scenarios (1997$ for IGSM, 2000$ for MERGE and MiniCAM, MER). U.S. economic growth is driven, in part, by labor force growth and, in part, by assumptions about productivity growth of labor and, in part, by other factors such as by savings and investment. Annual average growth rates are 1.4% for the MERGE reference scenario, 1.7% for the MiniCAM reference scenario, and 2.2% for the IGSM reference scenario. By comparison, U.S. GDP grew at an annual average rate of 3.4% from 1959-2004 (CEA 2005).

(MERGE) at roughly half of that of the highest scenario (IGSM) by the end of the century. The labor productivity growth assumptions for the U.S. in the IGSM scenario are the highest of the three, and the U.S. population assumptions are also relatively high in the IGSM scenarios. The relatively lower labor productivity growth assumptions used in the MERGE and MiniCAM scenarios lead to lower levels of GDP. The lower population growth assumptions employed in the MERGE scenarios give the MERGE reference scenario the lowest GDP in 2100.

Table 3.2 shows GDP across regions in the three reference scenarios. Differences in the absolute levels of GDP increase result from relatively small differences in rates of per capita growth. Although difficulties arise in comparisons of GDP across countries (see Box 3.1), the growth rates underlying these scenarios are usefully compared with historical experience. Long-term growth rates developed from reconstructed data (Table 3.3) show that consistent rapid growth is a phenomenon of industrialization, starting in the 1800s in North America and Europe and gradually spreading to other areas of the world. By the end of the period 1950 to 1973, it appeared that the phenomenon of rapid growth had taken hold in all major regions of the world. Since 1973, it has been less clear to what degree that conclusion holds. Growth slowed in the 1970s in most regions, the important exceptions being China, India, and several South and East Asian economies. In Africa, Latin America, Eastern Europe, and the former Soviet Union, growth slowed in this period to rates more associated with preindustrial times.

With this historical experience as background, the differences in GDP growth among the ref-

erence scenarios can be explained. Demographic trends, slowing population, and labor force growth all combine to influence overall GDP growth in the scenarios. With respect to the developed countries, the per capita income growth rate for the U.S. in the IGSM reference scenario is about the average for North America for the period 1950-2000. The MiniCAM reference scenario has lower growth, reflecting an assumption that an aging population will lead to lower labor force participation, and the result of this demographic maturation is a lower future rate of per capita GDP growth compared to history. U.S. growth rates in the MERGE reference scenario are similar to those of MiniCAM reference scenario.

GDP growth patterns for Western Europe and Japan are similar to one another within reference scenarios but vary across models. The IGSM reference scenario follows the post World War II trend in per capita GDP growth, but the MiniCAM and MERGE scenarios anticipate a break from the trend with lower per capita growth in GDP as a consequence of changes in underlying demographic trends. As with the U.S., the MiniCAM reference scenario exhibits a decline in average labor force participation in other developed regions as populations age, resulting in lower growth in per capita GDP compared to the IGSM reference scenario. The GDP growth pattern in the MERGE reference scenario is similar to that of MiniCAM reference scenario.

GDP growth patterns for developing regions show greater differences from historical experience. Notably, all three modeling groups chose assumptions leading to consistent growth in many non- Organization for Economic Cooperation and Development (OECD) regions at rates

63

Table 3.2. GDP for Key Regions Across Reference Scenarios, 2000-2100. This table reports GDP for all regions of the globe, but accounts for inconsistency in regional aggregations across models. Note that while regions are generally comparable, slight differences exist in regional coverage, particularly in aggregate regions. Differences in 2000 arise from these differences as do differences in regional deflators and regional exchange rates. *(Note: IGSM is in $1997 and 1997 exchange rates; MERGE uses $1997 and 1997 exchange rates restated to $2000 by the ratio of U.S. GDP for 2000 in $1997 and $2000; MiniCAM is in $2000 and 2000 exchange rates.)*

IGSM GDP by Region (trillions of $1997, MER)

Region	2000	2020	2040	2060	2080	2100
U.S.	9.1	16.9	29.3	44.4	59.8	76.4
Western Europe	9.2	15.8	27.0	41.5	57.2	74.2
Japan	4.4	7.5	13.8	21.8	30.0	38.6
Former Soviet Union	0.6	1.4	2.9	4.8	7.2	10.2
Eastern Europe	0.3	0.6	1.2	2.1	3.3	4.9
China	1.2	3.3	6.9	12.8	19.9	28.9
India	0.5	1.1	2.0	3.3	5.2	8.0
Africa	0.6	1.3	2.0	3.3	5.0	7.4
Latin America	1.6	3.0	6.3	11.5	18.0	25.9
Rest of the World	4.4	8.6	14.9	23.9	35.3	49.9

MERGE GDP by Region (trillions of $2000, MER)

Region	2000	2020	2040	2060	2080	2100
U.S.	9.8	16.1	20.9	26.8	33.1	39.6
Western Europe	9.8	14.4	19.9	26.9	35.0	43.6
Japan	4.6	6.0	7.7	9.6	11.7	13.9
Former Soviet Union/Eastern Europe	1.0	1.9	3.6	6.6	11.9	20.4
China	1.2	3.1	7.4	17.3	38.5	78.6
India	0.5	1.5	3.6	8.3	18.5	39.2
Africa/Latin America/Rest of World	6.5	14.6	27.5	49.3	85.1	141.9

MiniCAM GDP by Region (trillions of $2000, MER)

Region	2000	2020	2040	2060	2080	2100
U.S.	9.8	15.1	21.1	28.8	38.9	52.6
Western Europe	8.6	11.1	13.3	16.1	19.4	23.7
Japan	4.7	5.9	7.1	8.6	10.2	12.0
Former Soviet Union	0.4	0.8	1.4	2.3	3.6	5.7
Eastern Europe	0.4	0.7	1.4	2.4	4.0	6.6
China	1.2	4.8	11.6	20.8	34.1	49.3
India	0.5	1.6	4.8	10.7	19.5	32.0
Africa	0.6	1.2	2.1	3.9	7.7	13.8
Latin America	2.0	3.3	5.0	8.8	16.1	26.9
Rest of World	3.2	6.3	12.5	22.6	37.4	56.6

experienced by industrializing countries. However, growth rates are not homogeneous. Growth in China and India is generally higher than for regions such as Latin America and Africa, as it has been in recent decades. The IGSM reference scenario shows somewhat less growth for the non-OECD regions compared to the MiniCAM and MERGE reference scenarios. These are just one set of possible economic assumptions from each modeling group and are not intended to be expressions of what the groups view as desirable performance. Clearly, more rapid growth in developing countries, if gains spread to lower income groups within these regions, could be the basis for improving the outlook for people in these areas.

ENERGY USE, PRICES, AND TECHNOLOGY

The Evolving Structure of Energy Use

In the reference scenarios, global primary energy consumption expands dramatically over the century, growing to between 3 and

Region	1500-1820	1820-1870	1870-1913	1913-1950	1950-1973	1973-2001
North America	0.34	1.41	1.81	1.56	2.45	1.84
Western Europe	0.14	0.98	1.33	0.76	4.05	1.88
Japan	0.09	0.19	1.48	0.88	8.06	2.14
Former U.S.SR	0.10	0.63	1.06	1.76	3.35	-0.96
Eastern Europe	0.10	0.63	1.39	0.60	3.81	0.68
Africa	0.00	0.35	0.57	0.92	2.00	0.19
Latin America	0.16	-0.03	1.82	1.43	2.58	0.91
China	0.00	-0.25	0.10	-0.62	2.86	5.32
India	-0.01	0.00	0.54	-0.22	1.40	3.01
Other Asia	0.01	0.19	0.74	0.13	3.51	2.42

Table 3.3. Historical Annual Average per Capita GDP Growth Rates. *Source: Maddison, 2001*

BOX 3.1 Exchange Rates and Comparisons of Real Income Among Countries

Models used in this type of research typically represent the economy in real terms, following the common assumption that inflation is a purely monetary phenomenon that does not have real effects, but issues occur in comparing income across regions in terms of what currency exchange rates are most appropriate. The models do not represent the factors that govern exchange-rate determination and, therefore, cannot represent changes. However, modeling international trade in goods requires either an exchange rate or a common currency. Rather than separately model economies in native currencies and use a fixed exchange to convert currencies for trade, the equivalent and simpler approach is to convert all regions to a common currency at average market exchange rates (MER) for the base year of the model.

At the same time, it is widely recognized that using market exchange rates to compare countries can have peculiar implications. Country A might start with a larger GDP than country B when converted to a common currency using that year's exchange rates, and grow faster in real terms than B, yet could later have a lower GDP than B using exchange rates in that year. This paradoxical situation can occur if A's currency depreciates relative to B's. Depreciation and appreciation of currencies by 20% to 50% over just a few years is common, so the example is not extreme. Interest in making cross-country comparisons that are not subject to such peculiarities has led to development of indices of international purchasing power. A widely used index is purchasing power parity (PPP), whose development was sponsored by the World Bank. PPP-type indices have the advantage of being more stable over time and are thought to better reflect relative living standards among countries than MER. Thus, analysts drawing comparisons among countries have found it preferable to use PPP-type indices rather than MER. Although the empirical foundation for the indices has been improving, the theory for them remains incomplete, and thus there is a limited basis on which scenarios of future changes in PPP can be developed. Some hypothesize that differences close as real income gaps narrow, but the evidence for this outcome is weak, in part due to data limitations.

Controversy regarding the use of MER arose around the SRES produced by the IPCC (Nakicenovic et al. 2001) because they were reported to model economic convergence among countries, yet reported economic attributes of the scenarios in MER. Assessing convergence implies a cross-country comparison, but that would only be strictly meaningful if MER measures were corrected for a country's real international purchasing power. In developing the scenarios for this research, no assumptions were made regarding convergence. Growth prospects and other parameters for the world's economies were assessed relative to their own historical performance. The models used in this research are simulated in MER, as this is consistent with modeling of trade in goods. To the extent GDP results are provided, international comparisons are to made with great caution; for example, even global GDP for an historical period will differ if exchange rates of different years are used.

4 times its 2000 level of roughly 400 EJ. This growth results from a combination of forces, including rising economic activity, increasing efficiency of energy use, and changes in energy consumption patterns. Growth in per capita energy consumption occurs despite a continuous decline in the energy intensity of economic activity. The improvement in energy intensity reflects, in part, assumptions of substantial technological change in all three reference scenarios.

In all three reference scenarios a range of fossil resources is available to supply the bulk of the world's increasing demand for energy. Fossil fuels provided almost 90% of the energy supply in the year 2000 and remain the dominant energy source in all three reference scenarios throughout the twenty-first century despite a phasing out of conventional petroleum resources. Differing among the reference scenarios, however, is the mix of fossil fuels. The IGSM reference scenario has relatively more oil, derived from shale; the MERGE reference scenario has relatively more coal with a substantial amount of the increase used to produce liquid fuels; and the MiniCAM reference scenario has relatively more natural gas.

In all three reference scenarios, non-fossil fuel energy use grows substantially, reaching levels in 2100 that range from around half to levels that exceed total global energy consumption in 2000. The reference scenarios differ in terms of the mix of non-fossil resources. The substantial growth in non-fossil fuel energy use does not forestall substantial growth in fossil fuel consumption.

Energy production and consumption are closely associated with emissions of GHGs, particularly CO_2, because of the dominant role of fossil fuels in the energy sector. Figure 3.3 shows global primary energy consumption over the century and its composition by fuel type in the three reference scenarios. Not surprisingly, given the assumptions about economic growth, primary energy consumption grows substantially in all of the reference scenarios: from approximately 400 EJ/yr in the year 2000 to roughly between 1275 EJ/yr and 1500 EJ/yr by the end of this

century. Combined with population growth, all three reference scenarios include a growing per capita use of energy for the world (Figure 3.4). The per capita growth in primary energy consumption for the world is very similar for Mini-CAM and IGSM reference scenarios, with trends diverging somewhat late in the century. The MERGE reference scenario has relatively slower growth in per capita primary energy consumption early in the century, with accelerated growth later. On the other hand, per capita primary energy consumption in the U.S. differs substantially among the reference scenarios. U.S. per capita primary energy consumption in MERGE and IGSM reference scenarios increases substantially, while it declines gradually over the century in the MiniCAM reference scenario.

The growth in total and per capita primary energy consumption arises despite substantial improvements in energy technology assumed in all three scenarios. The ratio of U.S. primary energy consumption to GDP (primary energy intensity) declines throughout the century in all three reference scenarios (Figure 3.5). These patterns represent a continuation of changes in primary energy intensity that have occurred in recent decades in the U.S. In 2100, each dollar of real GDP is produced with only 40% of the primary energy consumed in 2000 in the MERGE reference scenario, only 30% of the energy in the IGSM reference scenario, and only 25% in the MiniCAM reference scenario.

Globally and in the U.S., primary energy consumption over the century remains dominated by fossil fuels. In this sense, the three reference scenarios tell a consistent story about future global energy, and all three run counter to the view that the world is running out of fossil fuels. Although reserves and resources of conventional oil and gas are limited in all three reference scenarios, the same cannot be said of coal and unconventional liquids and gases. In all three reference scenarios, the world economy moves from current conventional fossil resources to increased exploitation of some combination of the extensive (if more costly) global resources of heavy oils, tar sands, and shale oil, and to synfuels derived from coal. The three reference scenarios exhibit a different mix of these sources. The IGSM reference scenario exhibits

Figure 3.3. Global and U.S. Primary Energy Consumption by Fuel Across Reference Scenarios (EJ/yr). Global total primary energy consumption grows to between three and four times today's levels over the century in the reference scenarios, while U.S. primary energy consumption grows to between 1 and 2½ times today's levels. Fossil fuels remain a major energy source, despite substantial increases in the consumption of non-fossil energy sources. *[Notes. i. Oil consumption includes that derived from tar sands and oil shales, and coal consumption includes that used to produce synthetic liquid and gaseous fuels. ii. Primary energy consumption from nuclear power and non-biomass renewable electricity are accounted for at the average efficiency of fossil-fired electric facilities, which vary over time and across scenarios. This long-standing convention means that, all other things being equal, increasing efficiency of fossil-electric energy lowers the contribution to primary energy from these sources.]*

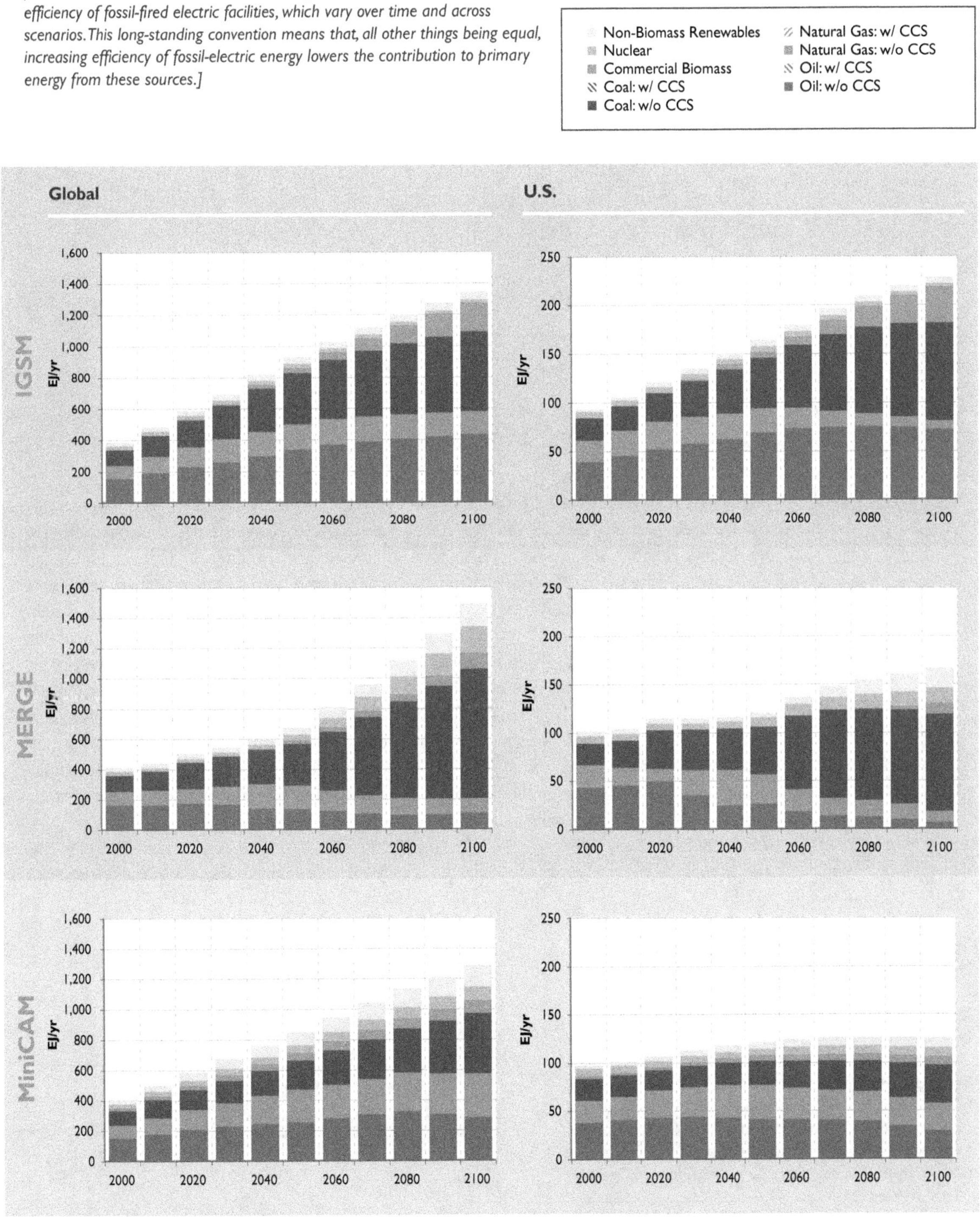

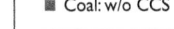

░ Non-Biomass Renewables	▨ Natural Gas: w/ CCS
▓ Nuclear	▨ Natural Gas: w/o CCS
▓ Commercial Biomass	▨ Oil: w/ CCS
▨ Coal: w/ CCS	■ Oil: w/o CCS
■ Coal: w/o CCS	

Figure 3.4. Global and U.S. Primary Energy Consumption per Capita Across Reference Scenarios (GJ per capita). All three reference scenarios include growing global per capita primary energy consumption. However, even after 100 years of growth, global per capita primary energy consumption is about ½ of the current U.S. level. U.S. per capita primary energy consumption varies more substantially among the reference scenarios. *[Note. Primary energy consumption from nuclear power and non-biomass renewable electricity are accounted for at the average efficiency of fossil-fired electric facilities, which vary over time and across scenarios.]*

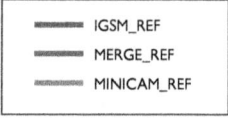

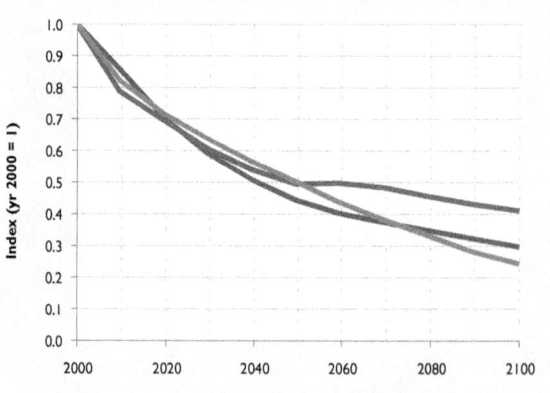

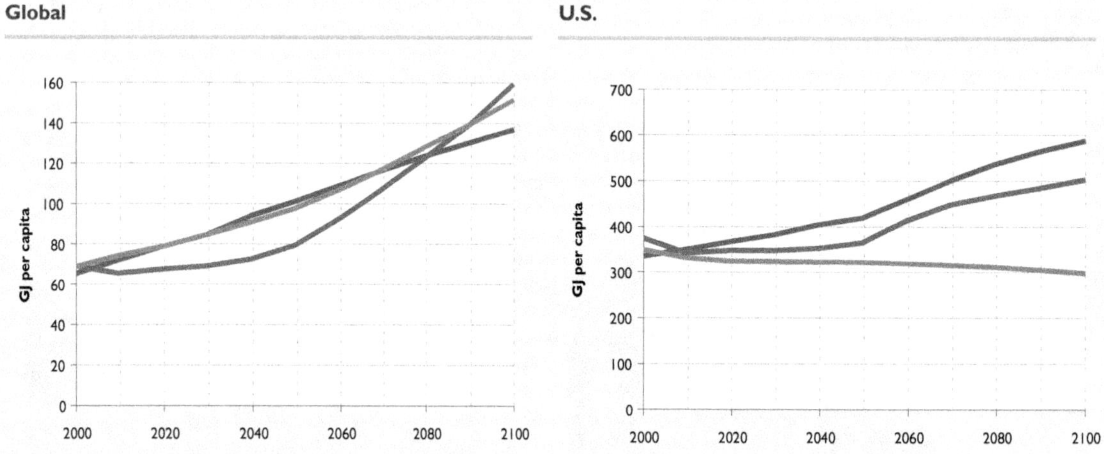

a relatively higher share of oil production (including unconventional oil); the MERGE reference scenario exhibits a relatively higher coal share; and the MiniCAM reference scenario exhibits a higher share for natural gas.

The relative contribution of oil to primary energy supply differs across the reference scenarios, but all three include a decline in the share of conventional oil. Thus, these scenarios represent three variations on a theme of energy transition precipitated by limited availability of conventional oil and continued expansion of final demands for liquid fuels, mainly for passenger and freight transport.

In the IGSM reference scenario, limits on the availability of conventional oil resources lead to the development of technologies to exploit unconventional oil, such as oil sands, heavy oils, and shale oil. These resources are large and impose no meaningful constraint on production during the twenty-first century. Thus, despite the fact that production costs are higher than for conventional oil, total oil production (conventional plus shale) expands throughout the century, although oil as a primary energy source declines as a share of total energy with the passage of time.

The transition plays out differently in the MERGE reference scenario. Although it begins

Figure 3.5. U.S. Primary Energy Intensity: Consumption per Dollar of GDP Across Reference Scenarios (Index, yr 2000 = 1.0). U.S. total primary energy intensity – primary energy consumption per dollar of GDP – continues to decline in the reference scenarios. In recent decades, the rate of decline has been about 14% per decade. U.S. primary energy intensity declines about 12% per decade in the IGSM reference scenario, about 13% per decade in the MiniCAM reference scenario, and about 9% per decade in the MERGE reference scenario. *[Note. Primary energy consumption from nuclear power and non-biomass renewable electricity are accounted for at the average efficiency of fossil-fired electric facilities, which vary over time and across scenarios.]*

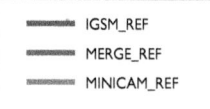

the same way (that is, the transition is initiated by limits on conventional oil resources), declining production of conventional oil leads to higher oil prices and makes alternative fuels, especially those derived from coal liquefaction, economically competitive. Thus, there is a transition away from conventional oil (and gas) and a corresponding expansion of coal production. The large difference between the MERGE and IGSM scenarios regarding primary oil thus reflects the role of coal liquefaction rather than a fundamentally different scenario of the need for liquid fuels.

The MiniCAM reference scenario depicts yet a third possible transition. Again, it begins with limited conventional oil resources leading to higher oil prices. Higher oil prices then lead to the development and deployment of technologies that access unconventional oil, such as oil sands, heavy oils, and shale oils. However, it also leads to expanded production of natural gas and to expanded production of coal to produce synthetic liquids, as in the MERGE reference scenario.

Primary energy consumption patterns also reflect assumptions about the availability of low-cost alternatives to conventional fossil fuels. In all three reference scenarios, non-fossil sources increase both their absolute and relative roles in providing energy to the global economy, with their share growing to roughly 20% to 30% of total supply by 2100. In the IGSM reference scenario, which has the lowest consumption from non-fossil resources, the magnitude of total consumption from these resources in 2100 is 65% the size of the total global primary energy consumption in 2000, which is more than a 500% increase in the level of production of non-fossil energy. In the MERGE reference scenario, which has the highest contribution from non-fossil resources, total primary energy consumption from these sources in 2100 exceeds total primary energy consumption in 2000. Despite this growth, the continued availability of relatively low-cost fossil energy supplies, combined with continued improvements in the efficiency with which they are used, allows fossil energy forms to remain competitive throughout the century.

The three reference scenarios tell different stories about non-fossil energy (much of which is covered below in the discussion of electricity generation). The IGSM reference scenario assumes political limits on the expansion of nuclear power, so it grows only to about 50% above the 2000 level by 2100. However, growing demands for energy and for liquid fuels in particular lead to the development and expansion of bioenergy, both absolutely and as percentage of total primary energy consumption.

In contrast, the MERGE reference scenario assumes that a new generation of nuclear technology becomes available and that societies do not limit its market penetration, so the share of nuclear power in the economy grows with time. In addition, renewable energy forms, both commercial biomass and other forms such as wind and solar, expand production during the century.

The MiniCAM reference scenario also assumes the availability of a new generation of nuclear energy technology that is both cost competitive and unrestrained by public policy. Nuclear power, therefore, increases market share although not to the extent found in the MERGE reference scenario. Non-biomass renewable energy supplies become increasingly competitive as well. In the MiniCAM reference scenario, bioenergy production is predominantly recycled wastes, with a modest contribution from commercial biomass farming toward the end of the century.

The three reference scenarios for the U.S. are similar in character to the global ones (Figure 3.3). The transition from conventional oil to alternative sources of liquid fuels and changes in electricity production affect energy markets and patterns in the U.S. However, primary energy consumption grows somewhat more slowly in the U.S. than in the world in general. As with the world total, the U.S. energy system remains dominated by fossil fuels in all three reference scenarios. The MERGE and IGSM reference scenarios have similar contributions from non-fossil energy, but the sources in the MERGE reference scenario are predominantly nuclear and other renewables, whereas it is biomass in the IGSM reference scenario. The MiniCAM reference scenario has the smallest overall contribution from non-fossil sources split relatively evenly between nuclear, biomass, and other renewables.

Trends in Fuel Prices

Historically, oil prices have been highly variable, with the volatility often related to political events (Figure 3.6). Prices were in the $15 to $20 range (in the constant 2006 dollars shown in the figure) until the increases in the 1970s and early 1980s that resulted from disruptions in the Middle East. In inflation-adjusted terms, prices declined from peaks in the late 1970s to vary around the $20 level in the latter half of the 1980s and 1990s. The period 2000 to 2005 has again seen rising prices of oil and other fossil energy sources, which suggests the possibility of a long-term trend toward rising prices. Depletion alone would suggest rising prices because of a combination of rents associated with a limited resource and the exhaustion of easily recoverable grades of oil. Global demand continues to grow, putting increasing pressure on supply. Improvements in technology that reduce the cost of recovering known deposits and facilitate discovery of new ones are opposing these forces toward higher prices.

The three models used for these scenarios employ time steps of 5 to 15 years (see Chapter 2) and, thus, are not set up to analyze short-term variability in prices. Their long-term trends are best interpreted as multi-year averages.

The three reference scenarios paint similar, but by no means identical, pictures of future energy prices. The price paths in the three reference scenarios reflect assumptions regarding both en-

ergy resources and energy technologies, and they shed light on these assumptions. For example, the price of oil is related to the marginal cost of bio-fuels, and the prices of both reflect, among other things, the technology options assumed to be available for their production.

Figure 3.7 shows mine-mouth coal prices, electricity producer prices, natural gas producer prices for the U.S., and the world oil (producer) price. All four energy markets – oil, natural gas, coal, and electricity – are shaped by the supply of, and demand for, these commodities. These fuels also are interconnected because users can substitute one fuel for another, thus higher prices in one fuel market will tend to increase demand for and the price of other fuels. Oil markets are driven by the rising cost of conventional oil and the transition to more expensive unconventional sources to supply a growing demand for liquid fuels, mainly for transportation. Thus, the oil prices in the scenarios result from the interplay between increasing demands for liquid fuels, the available technology, and the availability of liquids derived from these other sources.

Natural gas prices tell a similar story. Assumptions regarding the ultimately recoverable natural gas resource vary, as does the cost structure of the resource, leading to differences among the models. Like the demand for oil, the demand for natural gas grows, driven by increasing population and per capita incomes. As is the case for oil, the price of natural gas tends to be driven higher in the transition from inexpensive

Figure 3.6. Long-Term Historical Crude Oil Prices.
Crude oil prices have historically been highly variable. *(Figure courtesy of James Williams, WTRG Economics)*

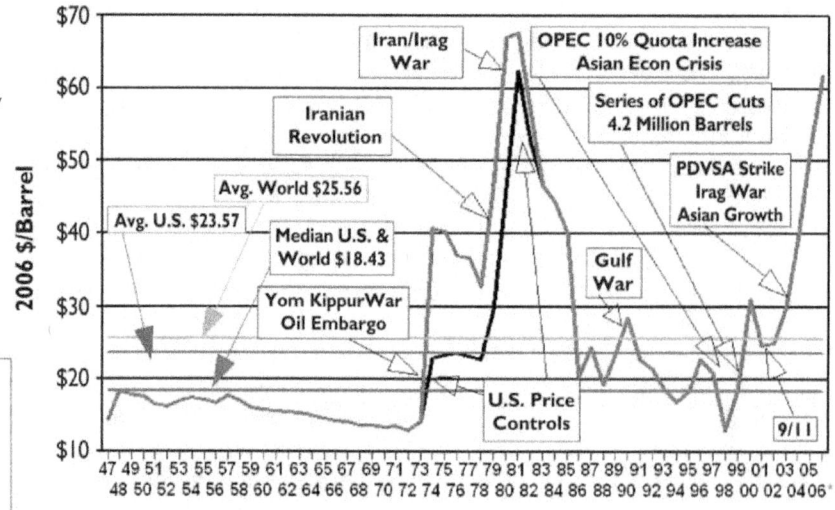

1947 – Sept. 2006

Figure 3.7. Indices of Energy Prices Across Reference Scenarios (Indexed to yr 2000 = 1.0). Energy prices through 2100 cover a wide range among the reference scenarios, but generally show a rising trend relative to recent decadal averages. Prices in the MERGE reference scenario are intermediate; by 2100 the crude oil price is about that observed in 2005 (3 times the 2000 level). The MiniCAM reference scenario has the lowest prices, with crude oil price about twice 2000 levels in 2100, somewhat below the level reached in 2005. The IGSM reference scenario has the highest prices, which for crude oil would be about 50% to 60% higher in 2100 than the price level of 2005.

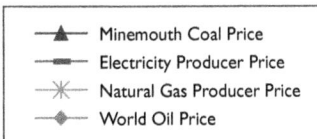

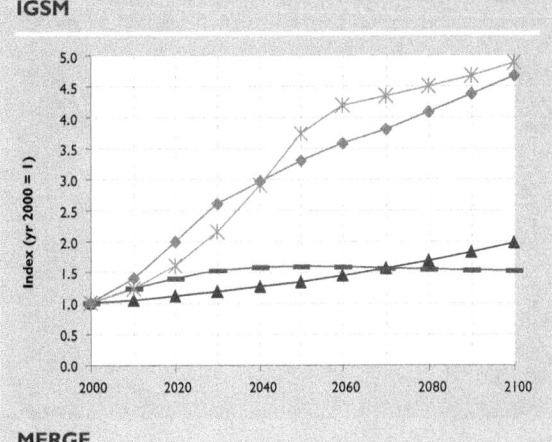

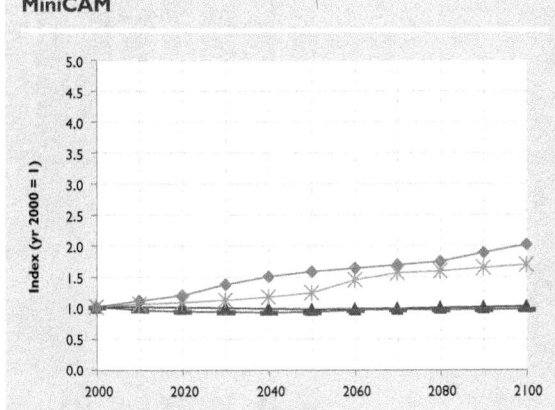

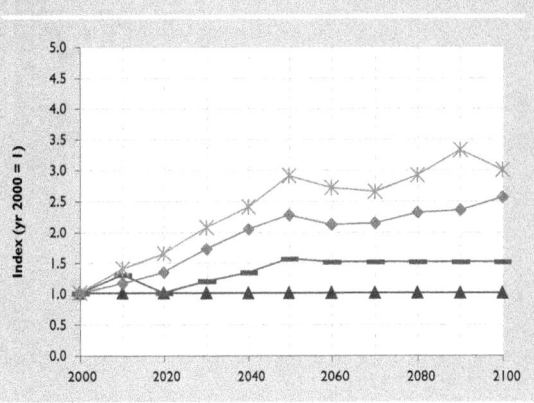

conventional resources to less easily accessible grades of the resource and to substitutes, such as gas derived from coal or biological sources. The different degrees and rates of price escalation reflect different technology assumptions in the three reference scenarios.

Coal prices do not rise as fast as oil and natural gas prices in any of the three reference scenarios. The reason is the abundance of the coal resource base. The different patterns of coal price movement with time in the three scenarios reflect differences in assumptions about the rate of resource depletion, its grade structure, and improvements in extraction technology.

The stability of electricity prices compared with oil and natural gas prices is a reflection of the variety of technologies and fuels available to produce electricity, their improvement over time, and the fact that fuel is just one component of the cost of electricity. The details un-

derlying this electric sector development are reported next.

Electricity Production and Technology

Electricity production steadily increases in both the U.S. and the world in the reference scenarios, although the scale and generation mix differ among the three reference scenarios (Figure 3.8). All the reference scenarios depict a continued role for coal. The IGSM reference scenario is dominated by coal, which accounts for more than half of all power production by the end of the twenty-first century. This characteristic of the IGSM reference scenario is consistent with its limited growth in nuclear power. In contrast, nuclear power penetrates the market based on economic performance, and non-biomass renewable energy gains market share in the MERGE reference scenario. Limited natural gas resources lead to a peak and decline in gas

Figure 3.8. Global and U.S. Electricity Production by Source Across Reference Scenarios (EJ/yr of electricity).
Global and U.S. electricity production in the reference scenarios show continued use of coal, especially the IGSM reference scenario, which assumes that nuclear energy expansion is limited by safety, waste, and proliferation concerns. The MERGE and MiniCAM reference scenarios are based on the assumption that nuclear energy is unconstrained by non-climate concerns, so these scenarios exhibit greater expansion. They also include greater contributions from renewable energy sources and somewhat greater use of electricity overall compared with IGSM reference scenario.

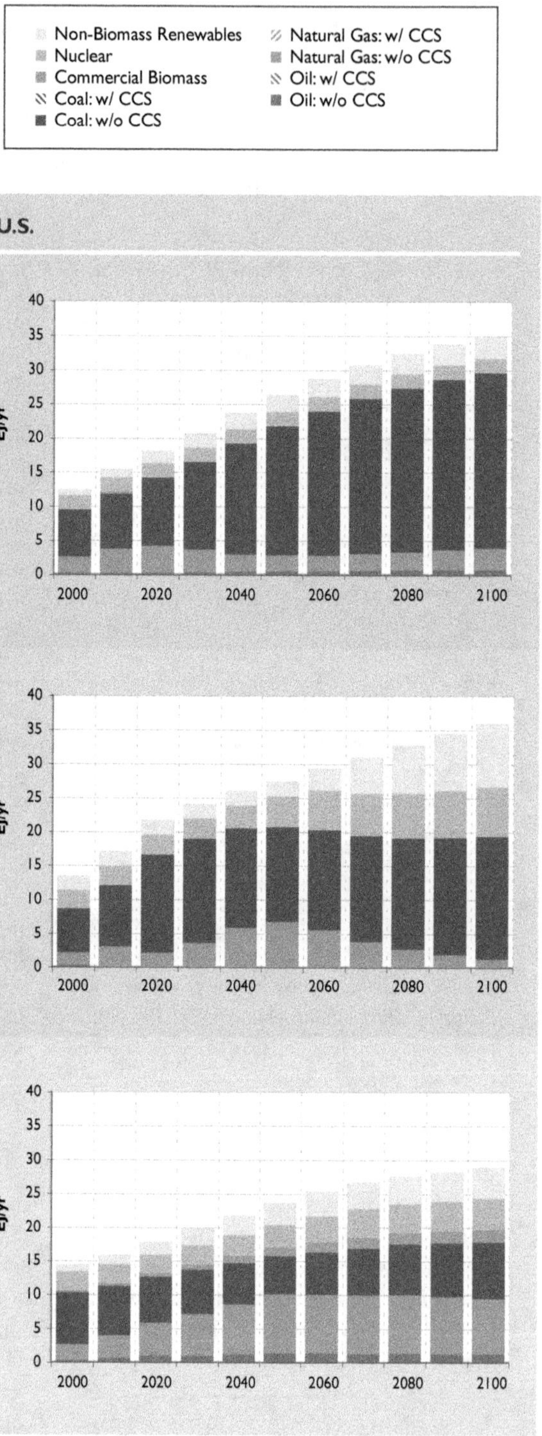

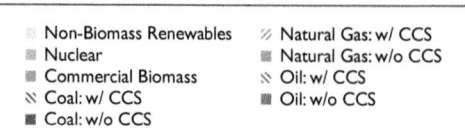

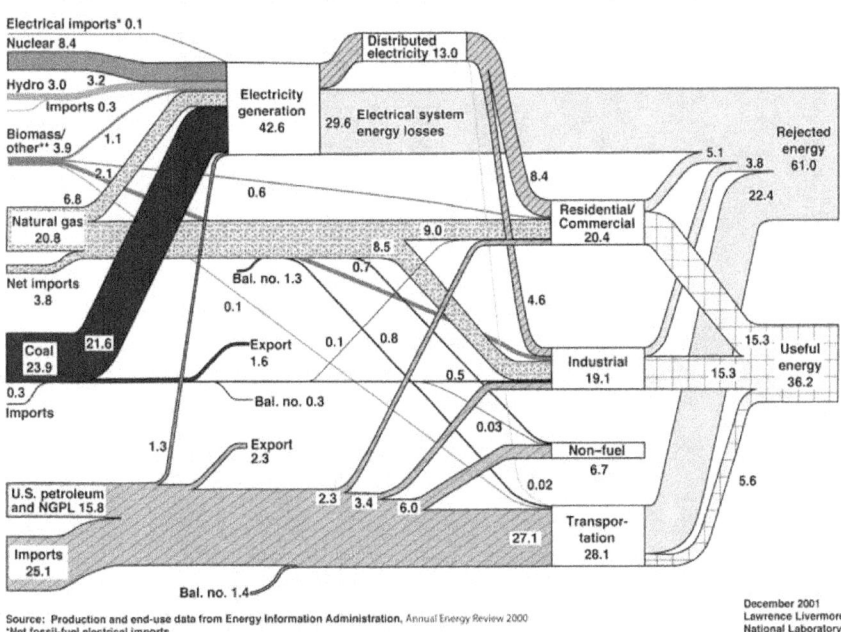

U.S. Energy Flow Trends – 2000
Net Primary Resource Consumption 104 Exajoules

Source: Production and end-use data from Energy Information Administration, *Annual Energy Review 2000*
*Net fossil-fuel electrical imports
**Biomass/other includes wood and waste, geothermal, solar, and wind.

December 2001
Lawrence Livermore
National Laboratory
http://eed.llnl.gov/flow

Figure 3.9. U.S. Energy Flow Diagram and Non-Electrical Energy Use for the Year 2000. Primary energy is transformed into different energy carriers that can easily be used for specific applications (e.g., space conditioning, light, and mechanical energy), but in the process losses occur. Of the 104 exajoules, of primary energy consumed in the U.S. in the year 2000, only an estimated 36 exajoules, were actually useful. Each of the models used in this research represents such conversion processes. Assumptions about efficiency improvements in conversion and end use are one of the reasons why energy intensity per dollar of GDP falls in the reference scenarios.

use in the first half of the century. In the Mini-CAM reference scenario, coal supplies the largest share of power, but natural gas is relatively abundant and provides a significant portion as well, as do nuclear and non-biomass renewable energy forms.

Non-Electric Energy Use

An important consideration in scenarios of the future energy system is conversion losses as relatively lower-grade resources are converted to higher-grade fuels for use in final applications such as space conditioning, lighting, and mechanical power. Figure 3.9 identifies the energy content of primary fuels for the U.S. in the year 2000 and where conversion losses occur. It shows the energy loss in the conversion from fuel to electricity to be 29.6 exajoules while the energy content of the electricity is 13.0 exajoules. Other losses occur when fuels are used to create the mechanical power to, for example, propel vehicles or when efficiency of conversion to heat, light, or mechanical energy is less that 100%. The potential for reducing such losses is one reason why energy intensity of the economy can continue to improve.

However, in the future other fuel transformation activities may become important and fundamentally change energy-flow patterns, as higher-grade resources are exhausted and lower-grade resources that require more conversion are used. As already discussed, the potential exists for coal and commercial biomass to be converted to liquids and gases – a technology thus far implemented only at a small scale. Furthermore, fuels and electricity may be transformed into hydrogen, creating fundamentally new branches of the energy system. Like electricity, these new branches will have conversion losses, and those losses can be important.

Figure 3.10 shows non-electric energy use in the reference scenario, and it is important to realize that these patterns of non-electric use also can imply significant conversion losses. This prospect plays a strong role in the MERGE reference scenario, in which coal and biomass go into liquefaction and gasification plants. To a lesser extent, these conversions are also present in the MiniCAM and IGSM scenarios. In addition, in the MiniCAM reference scenario some nuclear and renewable energy appears in non-electricity uses to produce hydrogen. In the IGSM and MiniCAM reference scenarios, oil

Figure 3.10. Global and U.S. Primary Energy Consumption in Non-Electric Applications Across Reference Scenarios (EJ/yr). As with electricity production, non-electric energy consumption remains heavily dependent on fossil fuels with some penetration of biomass energy. Primary energy is reported here, and the resurgence of coal in the reference scenarios is due to its use to produce synthetic liquids or gas. *[Notes. Oil consumption includes that derived from tar sands and oil shales, and coal consumption includes that used to produce synthetic liquid and gaseous fuels.]*

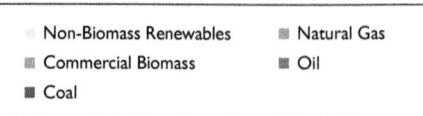

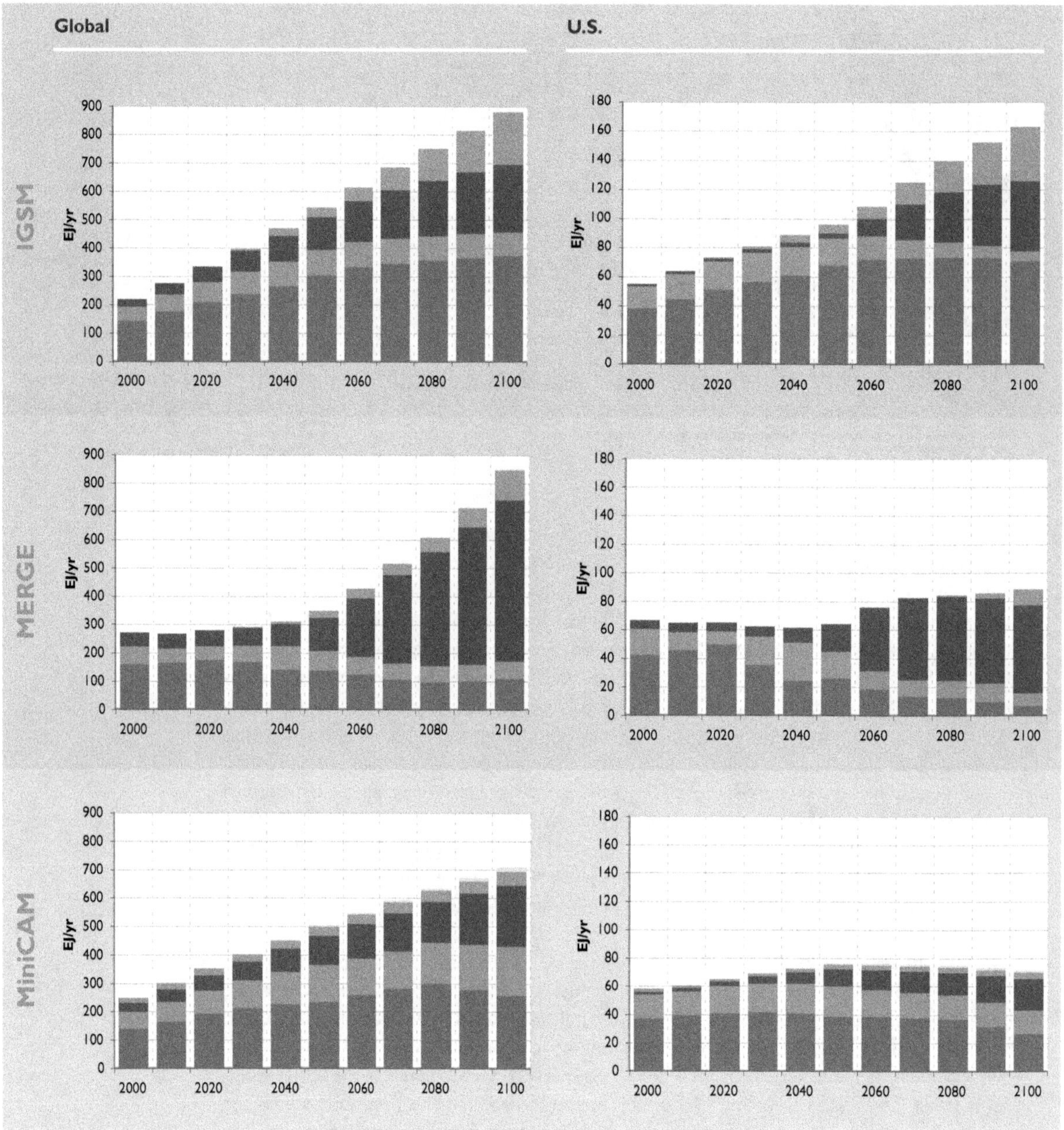

Figure 3.11. Global and U.S. Production of Biomass Energy Across Reference Scenarios (EJ/yr). The MiniCAM scenarios include waste-derived biomass fuels as well as commercial biomass and, thus, show significant use in 2000. The IGSM and MERGE scenarios include only commercial biomass energy beyond that already used. Globally, the IGSM and MERGE reference scenarios include more biomass production than does the MiniCAM reference scenario toward the end of the century.

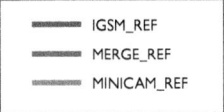

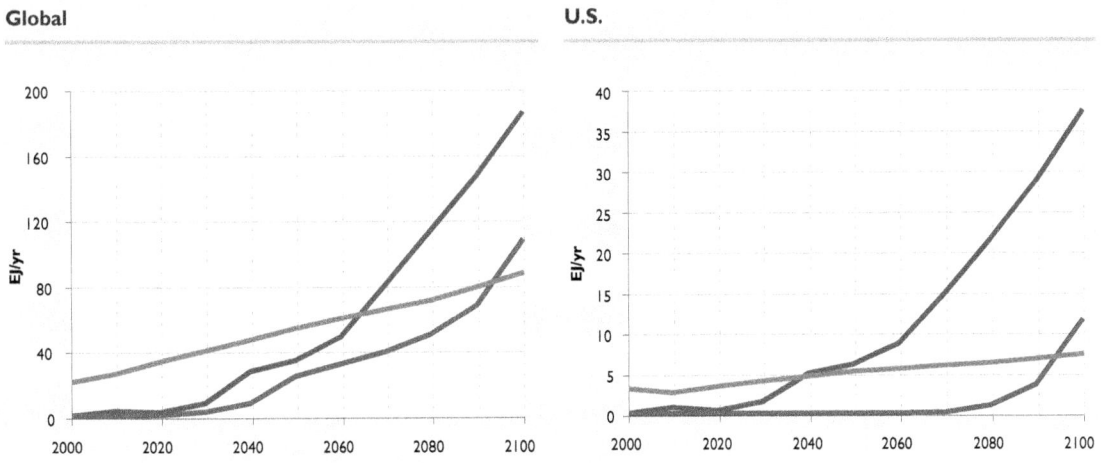

use is the largest single non-electric energy source, reflecting continued growth in demand for liquids by the transportation sectors. In the MERGE reference scenario, increasingly expensive conventional oil is supplanted by coal-based liquids. This phenomenon also has implications for energy intensity in that improvements in end-use energy intensity can be offset, in part, by losses in converting primary fuels to end-use liquids or gases.

LAND USE AND LAND-USE CHANGE

The three reference scenarios take different approaches to emissions from land use and land-use change. The MERGE reference scenario assumes that the biosphere makes no net contribution to the carbon cycle. In the IGSM and MiniCAM reference scenarios, the net contribution of the terrestrial biosphere is to remove carbon from the atmosphere, which results from the countervailing forces of land-use change emissions from deforestation and other human activities and the net uptake from unmanaged systems.

An important aspect of land use and land-use change in the scenarios from all three modeling

groups is the production of bio-fuels for energy. Both IGSM and MiniCAM take account of the competition for scarce land resources in developing scenarios of bioenergy production and consumption. MERGE takes the availability of bio-fuels as an exogenous input based on extra-model analysis. Global and U.S. biomass production is displayed in Figure 3.11. The IGSM and MiniCAM scenarios use somewhat different definitions, which account for the difference in 2000. The numbers presented for the IGSM scenarios account only the production of biomass energy beyond that now used and do not include traditional use of biomass or, for example, the own-use of wood wastes for energy in the forest products industry. The MiniCAM scenarios explicitly account for some current uses of biomass energy, such as that used in the pulp and paper industry, and separately consider the future potential for bio-fuels derived from wastes and residue along with energy crops grown explicitly for their energy content.

Apparent differences among the models need to be considered in light of this differential accounting. In the MiniCAM reference scenario, biomass production tends to be higher in early years because it is accounting waste and residue-derived bio-fuels explicitly. These waste and residue-derived bio-fuels account for all of the biomass production in the MiniCAM refer-

Figure 3.12. Global Net Emissions of CO$_2$ from Terrestrial Systems Including Net Deforestation Across Reference Scenarios (GtC/yr). Global net emissions of CO$_2$ from terrestrial systems, including net deforestation, serve as a slight net sink in 2000 that grows over time in the IGSM and MiniCAM reference scenarios, mainly because of reduced deforestation and CO$_2$ fertilization of plants. The MERGE scenarios assume a neutral terrestrial system.

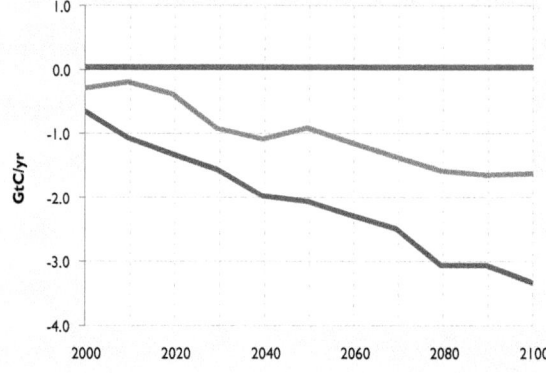

ence scenario in the early part of the century and the majority of all biomass production at the end of the century. The IGSM reference scenario exhibits strong growth in bio-fuels production beginning after the year 2020. Deployment in the IGSM reference scenario is driven primarily by a world oil price that in the year 2100 is over 4.5 times the price in the year 2000. In contrast, the MiniCAM reference scenario, with its lower long-term world oil price, includes insufficient incentive to create a substantial market for biomass crops. However, the MiniCAM reference scenario does include an increasing share of the potentially recoverable bio-waste as a source of energy.

Land use has implications for the carbon cycle as well. IGSM applies its component Terrestrial Ecosystem Model with a prescribed scenario of land use, and this land-use pattern is employed in all the IGSM scenarios. Thus, in the IGSM scenarios commercial biomass production must compete with other agricultural activities for cultivated land, but the extent of cultivated land does not change from scenario to scenario. Because the land-use pattern is fixed in the IGSM scenarios, changes in the net flux of carbon to the atmosphere reflect the behavior of the terrestrial ecosystem in response to changes in CO$_2$ and climatic effects that are considered within the IGSM's Earth system component. Taken together, these effects lead to the negative net emissions from the terrestrial ecosystem (Figure 3.12), which contrasts with the neutral biosphere assumed in the MERGE reference scenario. (Note that one tonne C is equivalent to 3.67 tonnes CO$_2$. See Box 3.2 for more on converting between units of carbon and units of CO$_2$.)

MiniCAM uses the terrestrial carbon cycle model of MAGICC (Wigley and Raper 2001, Wigley and Raper 2002) to determine the aggregate net carbon flux to the atmosphere. However, unlike either IGSM or MERGE, MiniCAM determines the level of terrestrial emissions as an output from an integrated agriculture-land-use module rather than as the product of a terrestrial model with fixed land use. Thus, the MiniCAM scenarios exhibit the same types of CO$_2$ fertilization effects as the IGSM scenarios, but they also represent interactions between the agriculture sector and the distribution of natural terrestrial carbon stocks.

EMISSIONS, CONCENTRATIONS, AND RADIATIVE FORCING

The growth in the global economy in the reference scenarios and the changes in the composition of the global energy system lead to growing emissions of GHGs over the century. Emissions from fossil fuel burning and cement production more than triple from 2000 to 2100 in all three reference scenarios. With growing emissions, GHG concentrations rise substantially over the twenty-first century, with CO$_2$ concentrations increasing by 2½ to over 3 times preindustrial levels. Increases in non-CO$_2$ GHG concentrations vary more widely across the reference scenarios. Radiative forcing from the GHGs considered in this research reaches 6.4 W/m^2 to 8.6 W/m^2 from preindustrial by 2100, with the non-CO$_2$ GHGs accounting for 20% to 25% of the instantaneous forcing in 2100.

Moderating the effect on the atmosphere of anthropogenic CO_2 emissions is the net uptake by the ocean and the terrestrial biosphere. As atmospheric CO_2 grows in the reference scenarios, the rate of net uptake by the ocean increases as well. Also, mainly through the effects of CO_2 fertilization, increasing atmospheric levels of CO_2 spur plant growth and net carbon uptake by the terrestrial biosphere. Differences among scenarios of these effects are, in part, a reflection of variation in their sub-models of the carbon cycle.

Greenhouse Gas Emissions

CALCULATING GREENHOUSE GAS EMISSIONS

Emissions of CO_2 from fossil fuels are the sum of emissions from each of the different fuel types, and for each type, emissions are the product of a fuel-specific emissions coefficient and the total combustion of that fuel. Exceptions to this treatment occur if a fossil fuel is used in a non-energy application (e.g., as a feedstock for plastic) or if the carbon is captured and stored in isolation from the atmosphere. All three of the modeling groups assumed the availability of CCS technologies and treated the leakage from such storage as zero over the time period considered in this research, although they assumed that carbon capture technologies capture and store less than 100% of the CO_2. CCS increases the costs of electricity production with no attendant benefits, absent actions to constrain carbon emissions, so CCS is not deployed in any of the reference scenarios.

Although bioenergy such as wood, organic waste, and straw are hydrocarbons like the fossil fuels (only much younger), they are treated as if their use had no net carbon release to the atmosphere. Any fossil fuels used in their cultivation, processing, transport, and refining are accounted for. Nuclear and non-biomass renewables, such as wind, solar, and hydroelectric power, have no direct CO_2 emissions and therefore have a zero carbon coefficient. Like bioenergy, emissions associated with the construction and operation of conversion facilities are accounted with the associated emitting source.

The calculation of net emissions from terrestrial ecosystems, including land-use change, is more complicated, and each model employs its own technique. IGSM employs the Terrestrial Ecosystem Model, which is a state-of-the-art terrestrial carbon-cycle model with a detailed, geographically disaggregated representation of terrestrial ecosystems and associated stocks and flows of carbon on the land. The IGSM scenarios, therefore, incorporate fluxes to the atmosphere as a dynamic response of managed and unmanaged terrestrial systems to the changes in the climate and atmospheric composition.

MiniCAM builds its net terrestrial carbon flux by summing both emissions from changes in the stocks of carbon from human-induced land-use change and the natural system response, represented in the reduced-form terrestrial carbon module of MAGICC. As noted above, Mini-CAM employs a simpler reduced-form representation of terrestrial carbon reservoirs and fluxes; however, its scenario is fully integrated with its agriculture and land-use module, which in turn is directly linked to energy and economic activity in the energy portion of the model. As noted above, the MERGE modeling group assumed no net emissions from the terrestrial biosphere.

Differing approaches among the modeling groups are used to account for the non-CO_2 GHGs. They begin with a current inventory of these gases and link growth in emissions to relevant activity levels. Because emissions are associated with very narrow activities, in some cases below the sectoral resolution of the models, emissions growth may be benchmarked to more detailed forecasts of activities.

REFERENCE SCENARIOS OF FOSSIL FUEL CO_2 EMISSIONS

All three reference scenarios include a transition from conventional oil production to some other source of liquid fuels based primarily on other fossil sources, either unconventional liquids or coal. As a consequence, carbon-to-energy ratios cease their historic pattern of decline (Figure 3.13). While the particulars of the reference scenarios differ, no reference scenario shows a dramatic reduction in carbon intensity over this century.

Figure 3.13. Global and U.S CO_2 Emissions from Fossil Fuel Combustion and Industrial Sources Relative to Primary Energy Consumption (GtC/EJ). The CO_2 intensity of energy use changes little over the century in the three reference scenarios, reflecting the fact that fossil fuels remain important sources of energy. Potential reductions in the CO_2 intensity of energy from more carbon-free or low-carbon energy sources is offset by a move to more carbon-intensive shale oil or synthetic fuels from coal.

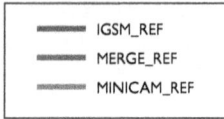

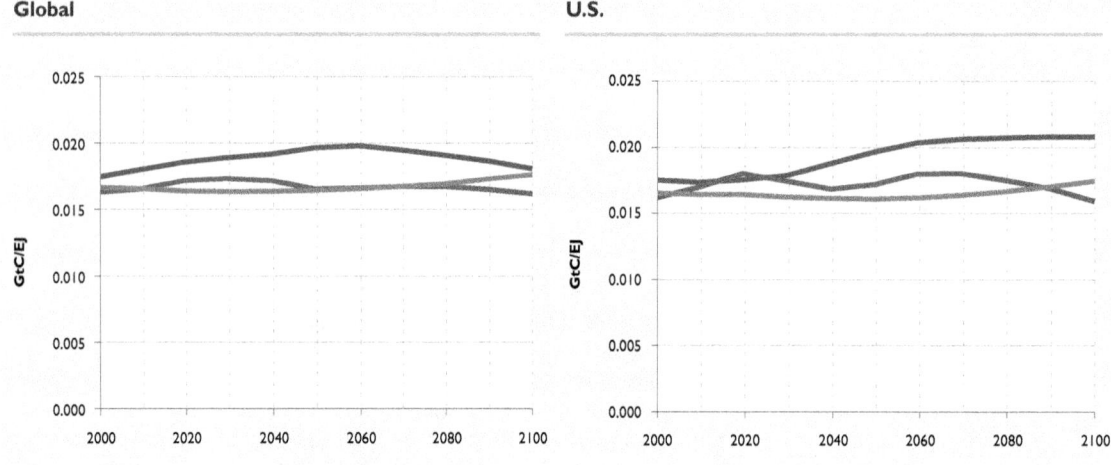

Substantial increases in total energy use with no or little decline in carbon intensity lead to substantial increases in CO_2 emissions per capita (Figure 3.14) and in global totals (Figure 3.15). Emissions of CO_2 from fossil fuel use and industrial processes increase from less than 7 GtC/yr in 2000 to between 22.5 and 24.0 GtC/yr by 2100. These global emissions are higher than in many earlier studies such as IS92a, where emissions were 20 GtC/yr in 2100 (Leggett et al. 1992). Global emissions from these reference scenarios are closer to those from the higher scenarios in the IPCC SRES (Nakicenovic et al. 2000); particularly those included under the headings A1FI and A2. U.S. emissions trajectories are more varied than the global trajectories. By 2100, U.S. emissions are between 2 GtC/yr and 5 GtC/yr.

The three reference scenarios display a larger share of emissions growth outside of the Annex I nations – the developed nations of the OECD as well as Eastern Europe and the former Soviet Union[1] (Figure 3.16). Annex I emissions are highest and Non-Annex I emissions lowest in the IGSM reference scenarios. At least in part, this is because of two factors underlying the

> **BOX 3.2   Reporting Conventions for Carbon Emissions and Prices**
>
> Two different conventions have been used to report emissions and prices in past studies of CO_2 emissions and concentrations. One convention is based on the total mass of emitted CO_2. Emissions are commonly expressed in tonnes of CO_2 and prices in terms of dollars per tonne of CO_2. The second convention is based on the carbon component of the emitted CO_2. Emissions are expressed in tonnes of carbon and prices in terms of dollars per tonne of carbon. This report uses the second approach throughout. In contrast, emissions of non-CO_2 GHGs, such as CH_4 or N_2O, are reported in terms of their full mass.
>
> It is important to be clear on which convention is used, but it is easy to convert between the two based on the molecular composition of CO_2. One molecule of CO_2 includes one carbon atom, with a molecular weight of 12, and two oxygen atoms, each with a molecular weight of 16. The total molecular weight of CO_2 is therefore 44, and carbon represents 12/44 of this weight. Emissions expressed in terms of CO_2 are therefore larger than when expressed in terms of the carbon component of CO2: one tonne of CO_2 is equivalent to 44/12, or 3.67, tonnes of carbon. Conversely, emissions prices are lower when reported in units of CO_2 because the price must be spread over a larger weight; $100 per tonne of carbon is equivalent to $27 per tonne of CO_2.

---

[1] Annex I is defined in the U.N. Framework Convention on Climate Change (FCCC [UN 1992]). However, since the FCCC entered into force, the Soviet Union has broken up. As a consequence, some of the republics of the former Soviet Union are now considered developing nations and do not have the same obligations as the Russian Federation under the FCCC. Thus, strictly speaking, the aggregations employed by the three modeling groups may not precisely align with the present partition of the world's nations. However, the quantitative implications of these differences are small.

Figure 3.14. Global and U.S. Emissions of CO$_2$ from Fossil and Other Industrial Sources per Capita Across Reference Scenarios (tonnes per capita). Global per capita fossil fuel and industrial CO$_2$ emissions grow in all three reference scenarios. However even after 100 years of growth, global per capita CO$_2$ emissions are slightly less than ½ of the 2000 U.S. level in the three scenarios. There is greater divergence in U.S. CO$_2$ emissions per capita over the century among the reference scenarios.

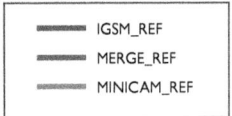

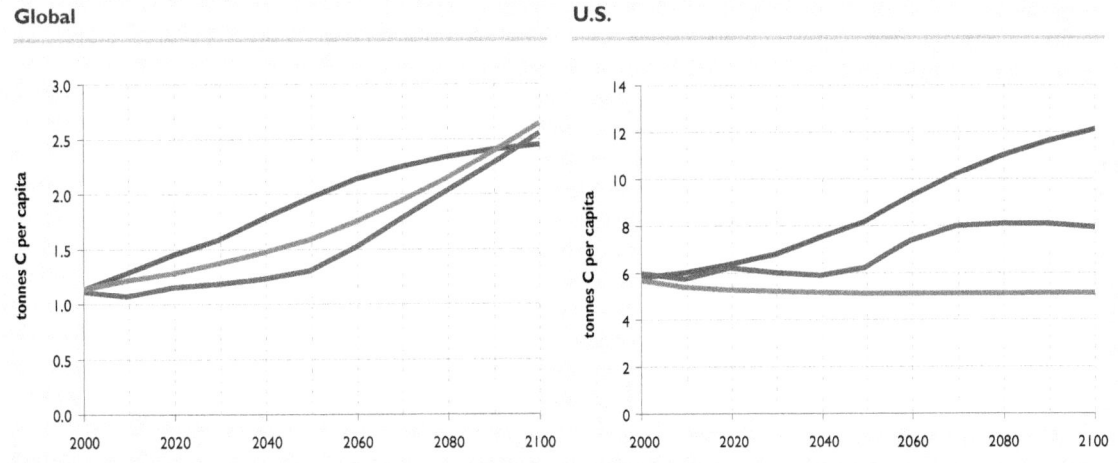

Figure 3.15. Global and U.S. Emissions of CO$_2$ from Fossil Fuels and Industrial Sources (CO$_2$ from land-use change excluded) Across Reference Scenarios (GtC/yr). Global emissions of CO$_2$ from fossil fuel combustion and other industrial sources, mainly cement production, grow throughout the century in all three reference scenarios. By 2100, global emissions are between 22.5 GtC/yr and 24.0 GtC/yr. U.S. emissions are more varied across the reference scenarios. By 2100, U.S. emissions are between 2 GtC/yr and 5 GtC/yr. Note that CO$_2$ from land-use change is excluded from this figure.

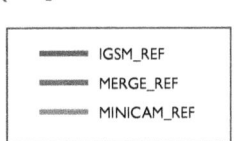

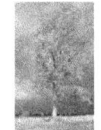

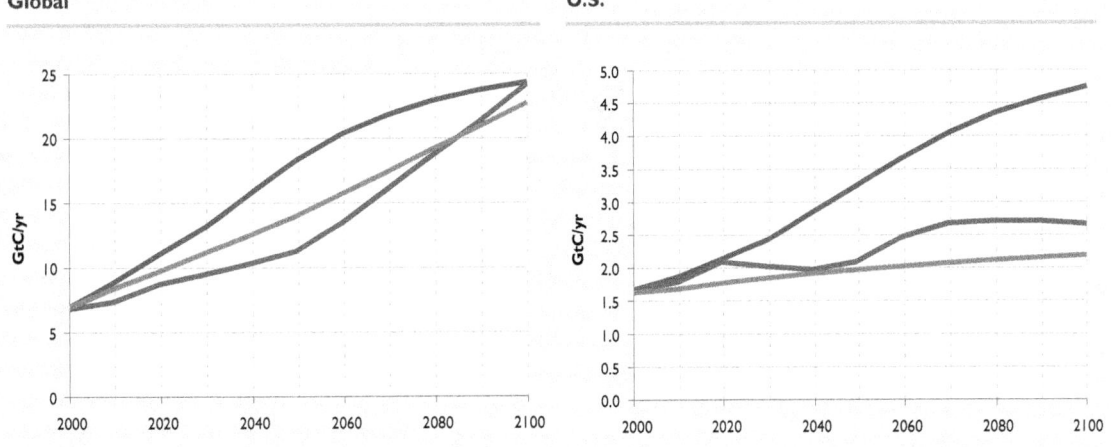

IGSM scenarios. First, the demand for liquids is satisfied by expanding production of unconventional oil, which has relatively high carbon emissions at the point of production. The U.S., with major resources of shale oil, switches from being an oil importer to an exporter but is responsible for CO$_2$ emissions associated with shale oil production. Second, assumed rates of

productivity growth in Non-Annex I nations are lower in the IGSM scenarios than in those of the other two models.

In contrast, the MERGE reference scenario assumes that liquids come primarily from coal, a fuel that is more broadly distributed around the world than unconventional oils. The MERGE

Figure 3.16. Global Emissions of Fossil Fuel and Industrial CO$_2$ by Annex I and Non-Annex I Countries Across Reference Scenarios (GtC/yr). Emissions of fossil fuel and industrial CO$_2$ in the Non-Annex I countries exceed Annex I emissions for all three reference scenarios by 2030 or earlier. The MERGE and MiniCAM reference scenarios exhibit continued relative rapid growth in emissions in Non-Annex I regions after that, so that emissions are on the order of twice the level of Annex I by 2100. The IGSM reference scenario does not show continued divergence, due in part to assumptions of relatively slower economic growth in Non-Annex I regions and faster growth in Annex I than the scenarios from the other modeling groups. The IGSM reference scenario also shows increased emissions in Annex I as those nations become producers and exporters of shale oil, tar sands, and synthetic fuels from coal

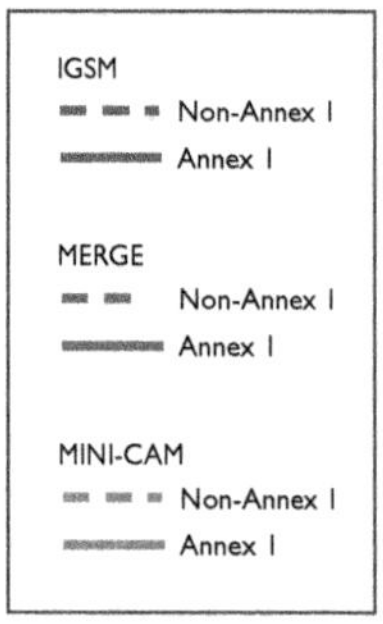

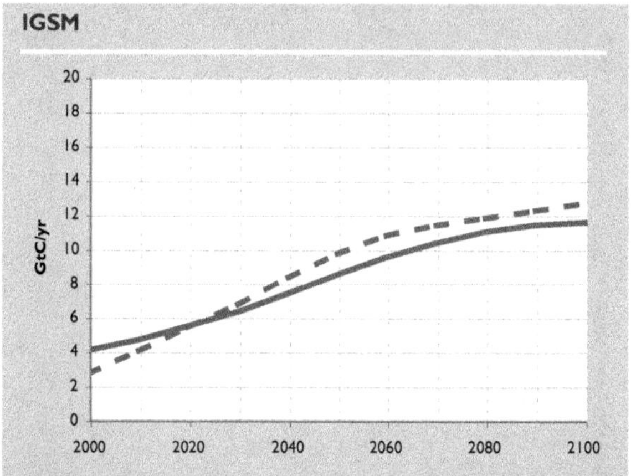

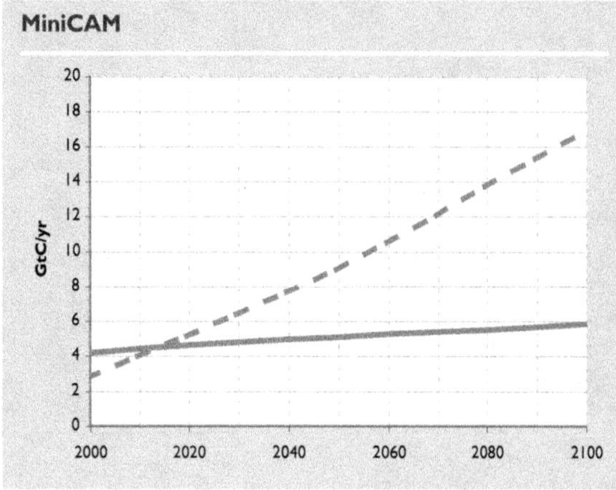

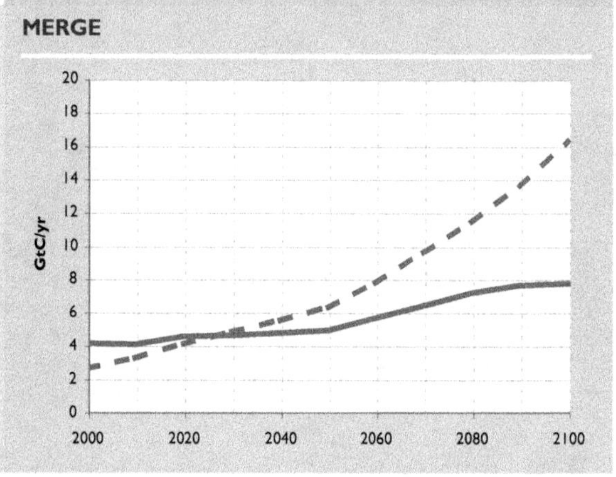

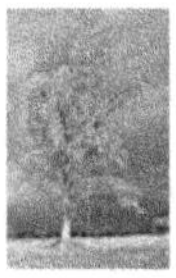

scenarios also exhibit higher rates of labor productivity in the Non-Annex I nations than the IGSM scenarios. Finally, the MERGE reference scenario has a greater deployment of nuclear power, leading to a lower carbon-to-energy ratio. These three features combine to produce lower Annex I emissions and higher Non-Annex I emissions than in the IGSM reference scenario. The MiniCAM reference scenario has Annex I emissions similar to those of the MERGE reference scenario, but higher Non-Annex I emissions.

The range of global fossil fuel and industrial CO$_2$ emissions across the three reference scenarios is relatively narrow compared with the uncertainty inherent in these developments over a century. While it is beyond the scope of this research to conduct a formal uncertainty or error analysis, both higher and lower emissions trajectories could be constructed.

There are at least two approaches to developing a sensible context in which to view these scenarios. One is to compare them with others produced by analysts who have taken on the same or a largely similar task. The literature on emissions scenarios is populated by hundreds of scenarios of future fossil fuel and industrial CO$_2$ emissions. Figure 3.17 gives some sense of what earlier efforts have produced, although they should be used with care. Many were developed at earlier times and may be significantly at variance with events as they have already unfolded. Also, no effort was undertaken in constructing the collection in the figure to weight scenarios for the quality of underlying analysis. Scenarios for which no underlying trajectories of population or GDP are available are mixed in with efforts that incorporate the combined wisdom of a large team of interdisciplinary researchers working over the course of years. Moreover, it is not clear that the observations are independent.

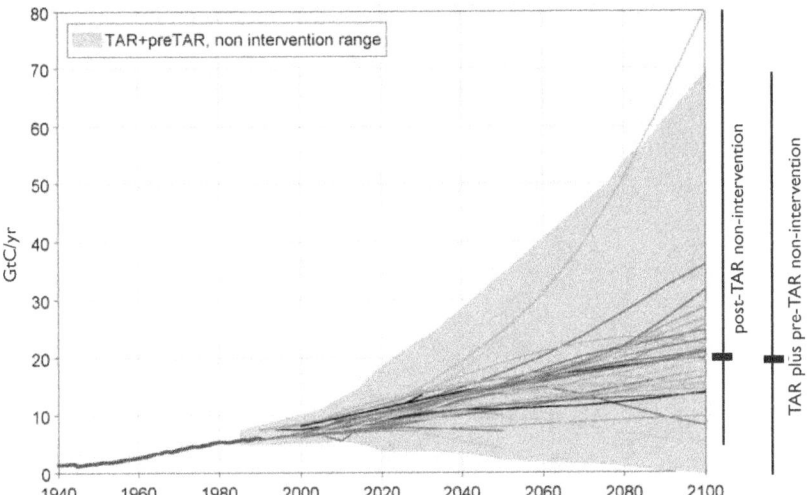

Figure 3.17. Global Emissions of CO$_2$ from Fossil Fuel and Industrial Sources: Historical Development and Scenarios (GtC/yr). The 284 non-intervention, or reference, scenarios published before 2001 are included in the figure as the blue-shaded range. The thin lines are an additional 55 non-intervention scenarios published since 2001. Two vertical bars on the right-hand side indicate the ranges for scenarios since 2001 (post-TAR non-intervention) and for those published up to 2001 (TAR plus pre-TAR non-intervention). *Source: Figure 4, Nakicenovic et al. 2006, with kind permission of Springer Science and Business Media.*

The clustering of year 2100 fossil fuel and industrial CO$_2$ emissions around 20 GtC/yr in both the pre- and post-IPCC TAR time frames coincides closely with the IPCC IS92a scenario. Many later scenarios were simply tuned to it, so are not independent assessments. For these reasons and others, looking to the open literature can provide some information, but caution in interpreting literature compilations is warranted.

Another approach to provide a context is systematic uncertainty analysis. There have now been several such analyses, including efforts by Nordhaus and Yohe (1983), Reilly et al. (1987), Manne and Richels (1994), Scott et al. (2000), and Webster et al. (2002). These studies contain many valuable lessons and insights. For the purposes of this research, one useful product of these uncertainty studies is an impression of the position of any one scenario within the window of futures that might pass a test of plausibility. Also useful is the way that the distribution of outcomes is skewed upward – an expected outcome when one considers that many model inputs, and indeed emissions themselves, are constrained to be greater than zero. Naturally, these uncertainty calculations present their own problems (Webster 2003).

FUTURE SCENARIOS OF ANTHROPOGENIC CH$_4$ AND N$_2$O EMISSIONS

The range of emissions for CH$_4$ and N$_2$O is wider than for CO$_2$ (Figure 3.18). Base-year emissions in the MERGE and MiniCAM reference scenarios are similar for N$_2$O but diverge for CH$_4$. In the IGSM reference scenario, CH$_4$ emissions are higher in the year 2000 than in the other scenarios, reflecting an independent assessment of historical emissions and uncertainty in the scientific literature regarding even historic emissions. Note that the IGSM reference scenario has a correspondingly lower natural CH$_4$ source (from wetlands and termites) that is not shown in Figure 3.18, balancing the observed concentration change, rate of oxidation, and natural and anthropogenic sources.

Both the IGSM and MERGE reference scenarios exhibit steadily growing CH$_4$ emissions throughout the twenty-first century as a consequence of the growth of CH$_4$-producing activities such as ruminant livestock herds, natural gas use, and landfills. Unlike CO$_2$, for which the combustion of fossil fuels without CCS leads inevitably to emissions, slight changes in activities can substantially reduce emissions of the non-CO$_2$ gases (Reilly et al. 2003). The MiniCAM reference scenario assumes that despite the expansion of human activities traditionally associated with CH$_4$ production, emissions control technologies will be deployed in response

Figure 3.18. Global CH₄ and N₂O Emissions Across Reference Scenarios (Mt CH₄/yr and Mt N₂O/yr). Global anthropogenic emissions of CH₄ and N₂O vary widely among the reference scenarios. There is uncertainty in year 2000 CH₄ emissions, with the IGSM reference scenario ascribing more of the emissions to human activity and less to natural sources. Differences in the scenarios reflect, to a large extent, different assumptions about whether current emissions rates will be reduced significantly for other reasons, for example, whether higher natural gas prices will stimulate capture of CH₄ for use as a fuel.

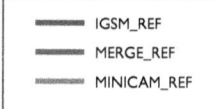

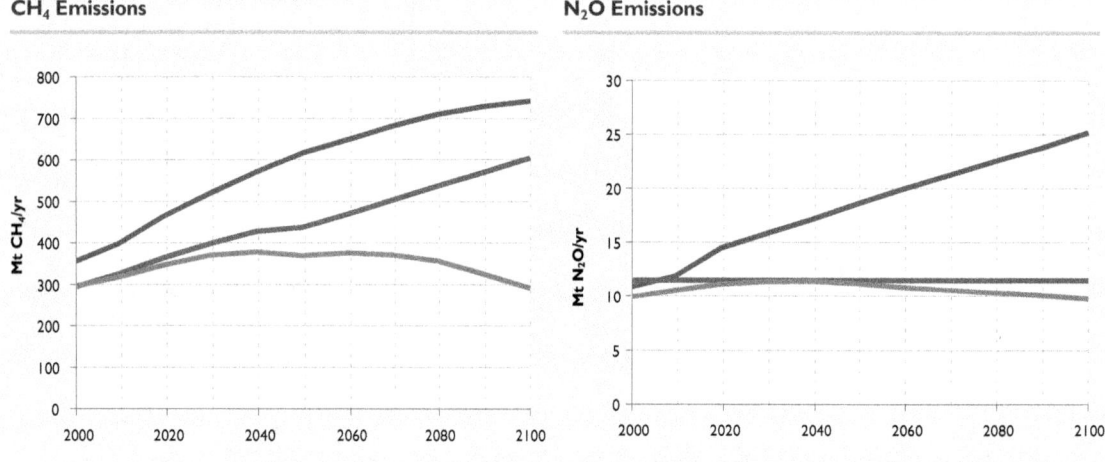

to local environmental regulations and in response to the economic value of CH₄. For this reason, CH₄ emissions peak and decline in the MiniCAM reference scenario.

FUTURE SCENARIOS OF ANTHROPOGENIC F-GAS EMISSIONS

A set of industrial products that act as GHGs are combined under the term, F-gases, which refers to an element that is common to them, fluorine. Several are replacements for the CFCs that have been phased out under the Montreal Protocol. They are usefully divided into two groups: (1) a group of HFCs, most of which are short-lived, and (2) the long-lived PFCs and SF₆. Figure 3.19 presents the reference scenarios for these GHGs. The IGSM and MERGE reference scenarios exhibit strong growth in the short-lived species, while the MiniCAM reference scenario exhibits about half as much growth over the century. Emissions of the long-lived gases are very similar among the reference scenarios. PFCs are used in semiconductor production and are emitted as a byproduct of aluminum smelting; they can be avoided relatively cheaply. Emissions from the main use of SF₆ in electric switchgear can easily be abated by recycling to minimize venting to the atmosphere. Many of the abatement activities have already been undertaken, and the modeling groups assumed they will continue to be used.

The Carbon Cycle: Net Ocean and Terrestrial CO₂ Uptake

The stock of carbon in the atmosphere at any time is determined from an initial concentration of CO₂ to which is added anthropogenic emissions from fossil fuel and industrial sources and from which is subtracted net CO₂ transfer from the atmosphere to the ocean and terrestrial systems. Each of the three participating models represents these processes differently.

The three reference scenarios display strong increases in ocean uptake of CO₂ (Figure 3.20), reflecting modeled mechanisms that become increasingly active as CO₂ accumulates in the atmosphere. The IGSM reference scenario has the least active ocean, which results from its three-dimensional ocean representation that shows less uptake, in part, as a result of rising water temperatures and CO₂ levels in the surface layer and, in part, as a result of a slowing of mixing into the deep ocean. The MERGE reference scenario has the most active ocean, and uptake rates continue to increase over the century. As will be discussed in Chapter 4, the three ocean models produce more similar behavior in the stabilization scenarios; for example, the MERGE and MiniCAM Level 1 and Level 2 scenarios have almost identical ocean uptake.

Figure 3.19. Global Emissions of Short-Lived and Long-Lived F-Gases (Kt HFC-134a-Equivalent/yr and Kt SF₆-Equivalent/yr).

Short-Lived F-Gas Emissions

Long-Lived F-Gas Emissions

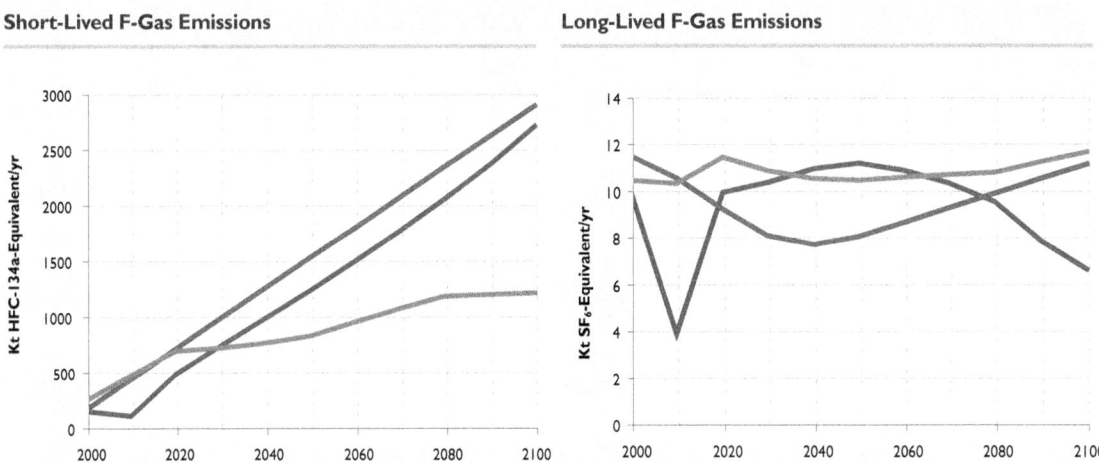

As discussed above, the net transfer of CO_2 from the atmosphere to terrestrial systems includes many processes, such as deforestation (which transfers carbon from the land to the atmosphere), uptake from forest regrowth, and the net effects of atmospheric CO_2 and climate conditions on vegetation. As noted earlier, MERGE employs a neutral biosphere: by assumption, its net uptake is zero with processes that store carbon assumed to just offset those that release it. Taken together with its more active ocean system in the reference scenario, the behavior of the carbon cycle in total is similar to the other two models, especially MiniCAM. IGSM and MiniCAM employ active terrestrial biospheres, which on balance remove carbon from the at-

mosphere (Figure 3.12). Both the MiniCAM and the IGSM reference scenarios display the net effects of deforestation, which declines in the second half of the century, combined with terrestrial processes that accumulate carbon in existing terrestrial reservoirs. The IGSM and MiniCAM reference scenarios also include feedback effects of a changing climate.

Greenhouse Gas Concentrations

Radiative forcing is related to the concentrations of GHGs in the atmosphere. The relationship between emissions and concentrations of GHGs is discussed in Box 2.2. The concentration of gases that reside in the atmosphere for long pe-

Figure 3.20. CO₂ Uptake from Oceans Across Reference Scenarios (GtC/yr, expressed in terms of net emissions). The IGSM reference scenario, which is based on the IGSM's three-dimensional ocean model, exhibits less CO_2 uptake than the other two reference scenarios and, after some point, little additional increase in uptake even though concentrations are rising.

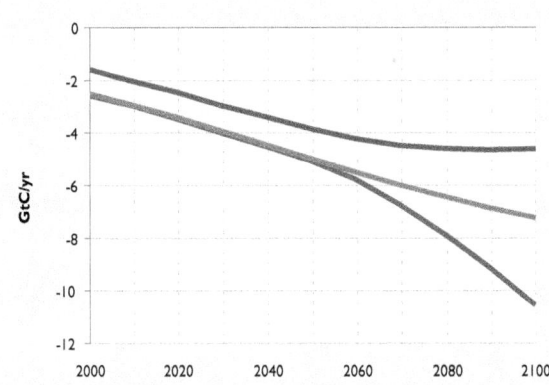

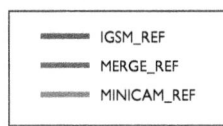

The MiniCAM reference scenario exhibits some slowing of ocean uptake, although not as pronounced as in the IGSM reference scenario. There is no slowing of uptake in the MERGE reference scenario. Although the MERGE reference scenario has higher ocean uptake in the latter half of the century, the effects of this increase are offset by the assumption of a neutral biosphere. Hence the aggregate behavior of its carbon cycle tends to be more similar to that in the other two reference scenarios, especially the MiniCAM reference scenario (Figure 3.22). The three ocean models produce more similar behavior in the stabilization scenarios.

Figure 3.21. Relationship Between Cumulative CO₂ Emissions from Fossil Fuel Combustion and Industrial Sources, 2000-2100, and Atmospheric CO₂ Concentration in 2100 Across All Scenarios.
Despite differences in how the carbon cycle is handled in each of the three models, the scenarios exhibit a very similar response in terms of concentration level for a given level of cumulative emissions. [Note. The cumulative emissions do not include emissions from land use and land-use change].

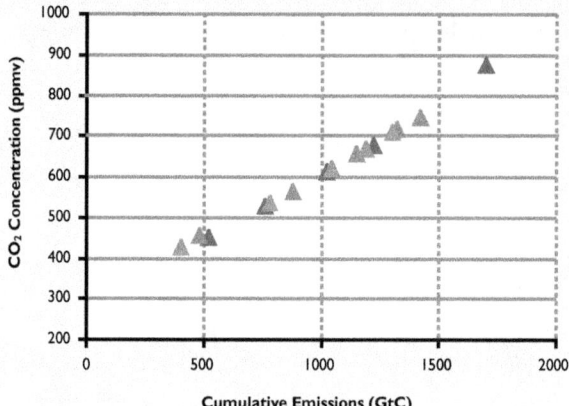

riods of time – decades to millennia – is more closely related to cumulative emissions than to annual emissions. In particular, this is true for CO_2, the gas responsible for the largest contribution to radiative forcing. This relationship can be seen for CO_2 in Figure 3.21, where cumulative emissions over the period 2000 to 2100, from the three reference scenarios and the twelve stabilization scenarios, are plotted against the CO_2 concentration in the year 2100. The plots for all three models lie on essentially the same line, indicating that despite considerable differences in representation of the processes that govern CO_2 uptake, the aggregate response to increased emissions is very similar. This basic linear relationship also holds for other long-lived gases, such as N_2O, SF_6, and the other long-lived F-gases.

GHG concentrations rise in all three reference scenarios. CO_2 concentrations increase from 370 ppmv in year 2000 to somewhere in the range of 700 to 875 ppmv in 2100 (Figure 3.22). The preindustrial concentration of CO_2 was approximately 280 ppmv. While all three reference scenarios display the same increasing pattern, by the year 2100 there is a difference of approximately 175 ppmv among the three scenarios. This difference has implications for radiative forcing and emissions mitigation (discussed in Chapter 4).

Increases in the concentrations of the non-CO_2 GHGs vary across the reference scenarios. The concentrations of CH_4 and N_2O in the MiniCAM reference scenario are on the low end of the range, reflecting assumptions discussed above about use of CH_4 for energy and emissions control for non-climate reasons. The

IGSM reference scenario has the highest concentrations for all of the substances. The differences mainly reflect differences in anthropogenic emissions, but they also are influenced by the way each model treats natural emissions and sinks for the gases. The IGSM scenarios include climate and atmospheric feedbacks to natural systems, which tend to result in an increase in natural emissions of CH_4 and N_2O. Also, increases in other pollutants generally lengthen the lifetime of CH_4 in the IGSM scenarios because the other pollutants deplete the atmosphere of the hydroxyl radical (OH), which is the removal mechanism for CH_4. These feedbacks tend to amplify the difference in anthropogenic emissions among the reference scenarios. The concentrations of the short-lived and long-lived F-gases are also presented in Figure 3.22.

Radiative Forcing from Greenhouse Gases

Contributions to radiative forcing are a combination of the abundance of the gas in the atmosphere and its heat-trapping potential (radiative efficiency). Of the directly released anthropogenic gases, CO_2 is the most abundant, measured in parts per million; the others are measured in parts per billion. However, the other GHGs are about 24 times (CH_4), to 200 times (N_2O), to thousands of times (SF_6 and PFCs) more radiatively efficient than CO_2. Thus, what they lack in abundance they make up for, in part, with radiative efficiency. However, CO_2 is still the main contributor to radiative forcing among these substances, and all three reference scenarios exhibit an increasing relative contribution from CO_2.

Figure 3.22. Atmospheric Concentrations of CO_2, CH_4, N_2O, and F-gases Across the Reference Scenarios (units vary). Differences in concentrations for CO_2, CH_4, and N_2O across the reference scenarios reflect differences in emissions and treatment of removal processes. By 2100, CO_2 concentrations range from about 700 ppmv to 900 ppmv, CH_4 concentrations range from 2000 ppbv to 4000 ppbv, and N_2O concentrations range from about 380 ppbv to 500 ppbv.

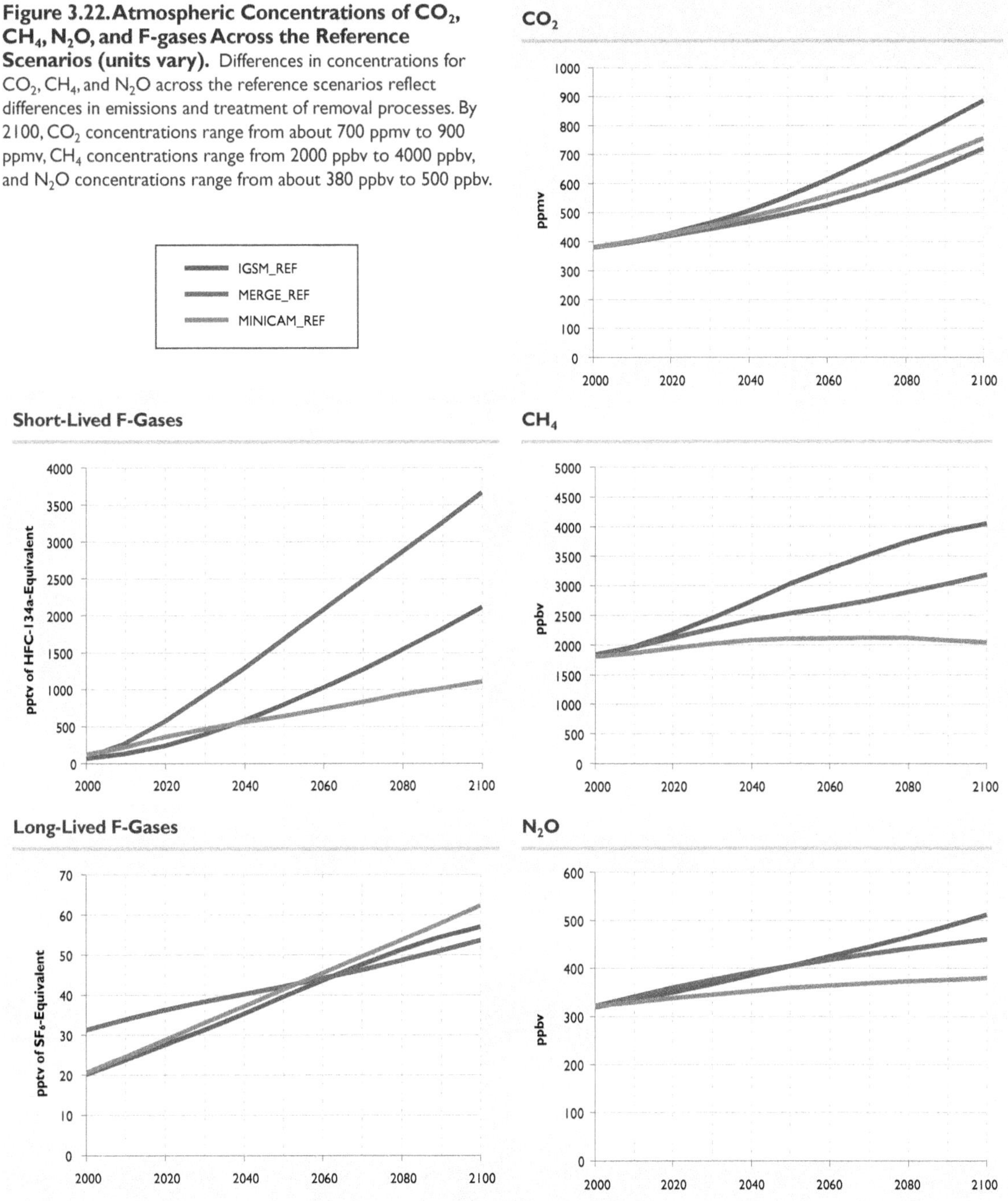

- IGSM_REF
- MERGE_REF
- MINICAM_REF

CO₂

Short-Lived F-Gases

CH₄

Long-Lived F-Gases

N₂O

The three models display essentially the same relationship between GHG concentrations and radiative forcing, so the three reference scenarios also all exhibit higher radiative forcing, growing from roughly 2.1 W/m² from preindustrial in 1998 to between 6.4 W/m² and 8.6 W/m² in 2100. The differences among radiative forcing in 2100 imply differences in the amount of emissions reductions required to stabilize as the four radiative forcing levels in this research. For example, the emissions reductions required for stabilization in the IGSM stabilization scenarios are larger than those required in the Mini-CAM stabilization scenarios, because the

Figure 3.23. Radiative Forcing by Gas Across Reference Scenarios (W/m² from preindustrial).
CO_2 accounts for more than 80% of the radiative forcing from the GHGs considered in this research by 2100 in all three reference scenarios. Radiative forcing in 2100 ranges from about 6.4 W/m² to 8.6 W/m² from preindustrial in the reference scenarios.

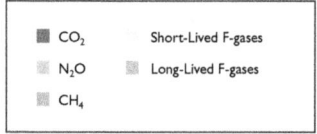

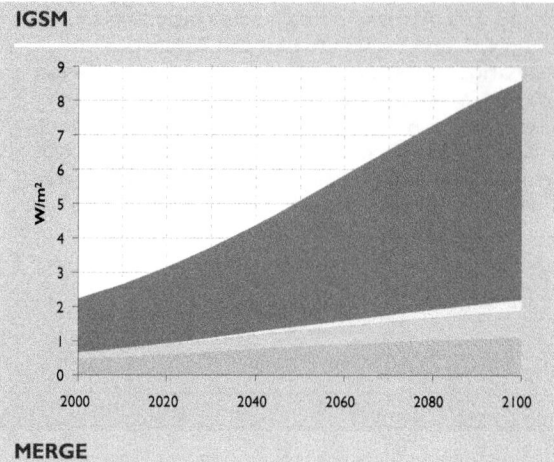

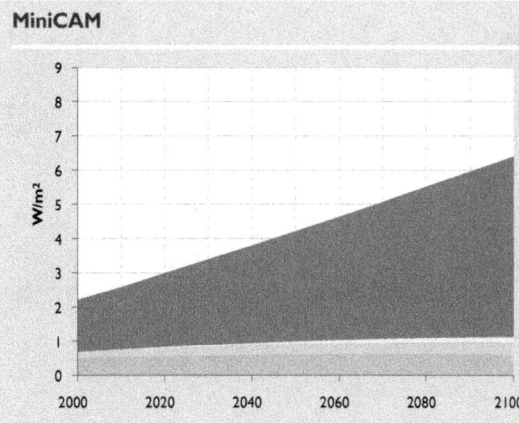

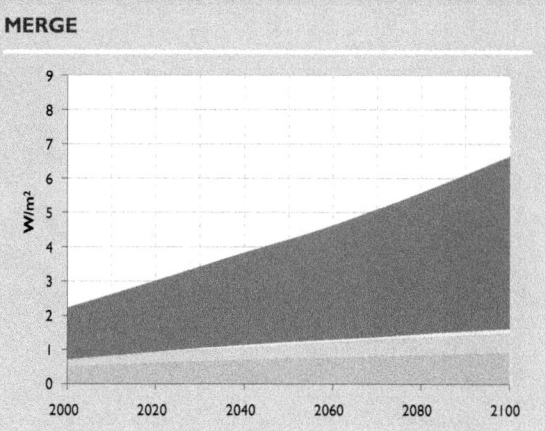

radiative forcing reaches 8.6 W/m² in 2100 in the IGSM reference scenario and 6.4 W/m² in the MiniCAM reference scenario.

The relative contribution of CO_2 to radiative forcing increases over the century in all three reference scenarios (Figure 3.23). In 2000, the non-CO_2 GHGs examined in this research contributed slightly above 30% of the estimated radiative forcing from preindustrial. In the IGSM reference scenario, the contribution of the non-CO_2 GHGs to radiative forcing falls slightly to about 26% by 2100. The MiniCAM reference scenario includes little additional increase in radiative forcing for non-CO_2 GHGs, largely as a result of assumptions regarding the control of CH_4 emissions for non-climate reasons, and thus has their share falling to about 18% by 2100. The MERGE reference scenario is intermediate, with the non-CO_2 GHG contribution falling to about 24%.

From the discussion above, it can be seen that the three reference scenarios contain many large-scale similarities. All have expanding global energy systems, all remain dominated by fossil fuel use throughout the twenty-first century, all generate increasing concentrations of GHGs, and all produce substantial increases in radiative forcing. Yet the reference scenarios differ in many details, ranging from demographics to labor productivity growth rates to the composition of energy supply to treatment of the carbon cycle. These differences shed light on important points of uncertainty that arise for the future. In Chapter 4, they will also be seen to have important implications for efforts to limit radiative forcing.

Stabilization Scenarios

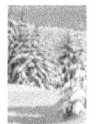

In these scenarios, stabilizing radiative forcing at levels ranging from 3.4 W/m² to 6.7 W/m² above preindustrial levels (Level 1 to Level 4) implies significant changes to the world's energy and agriculture systems and leads to lower global economic output. Although all the stabilization scenarios require changes in the world's energy and agricultural systems, the three modeling groups produced scenarios with differing conceptions of how these changes might occur. The economic implications vary considerably among the scenarios, depending on the amount that emissions must be reduced and the evolution of technology, particularly in the post-2050 period.

INTRODUCTION

In Chapter 3, each modeling group developed scenarios of long-term GHG emissions associated with changes in key characteristics, such as demographics, economic growth, and technology. This chapter describes how such developments might affect or be affected by limits on radiative forcing. It illustrates that society's response to a limit on radiative forcing can take many paths, reflecting factors shaping the reference scenario and the availability and performance of emissions-reducing technologies. Control of GHG emissions requires changes in the global energy, economic, agriculture, and land-use systems.

It should be emphasized that the four radiative forcing stabilization levels considered in this research and detailed in Table 1.2 were chosen for illustrative purposes only. They reflect neither a preference nor a recommendation. In all the stabilization scenarios, it was assumed that radiative forcing would not be allowed to overshoot the radiative forcing levels along the path to long-term stabilization. Given this assumption, each modeling group had to make further decisions regarding the means of meeting these radiative forcing limits. Section 4.2 compares the approaches of the three modeling groups. Section 4.3 shows the effect of the three strategies on GHG emissions, concentrations, and radiative forcing. The implications for global and U.S. energy and industrial systems are explored in Section 4.4 and for agriculture and land-use change in Section 4.5. Section 4.6 discusses economic consequences of the measures to achieve the various radiative forcing stabilization levels in these scenarios.

STABILIZING RADIATIVE FORCING: MODEL IMPLEMENTATIONS

Some features of scenario construction were coordinated among the three modeling groups, and others were left to their discretion. In three areas, a common set of approaches was adopted:

- Climate policies in the stabilization scenarios

- The timing of participation in stabilization scenarios

- Policy instrument assumptions in stabilization scenarios.

In two areas, the groups employed different approaches:

- The timing of CO_2 emissions mitigation

- Non-CO_2 emissions mitigation.

Climate Policies in the Stabilization Scenarios

For the stabilization scenarios, each modeling group assumed that, as in the reference scenarios, the U.S. will achieve its goal of reducing GHG emissions intensity (the ratio of GHG emissions to GDP) by 18% by 2012, although implementation of this goal was left to the judgment of each modeling group. Also, the Kyoto Protocol participants were assumed to achieve their commitments through the first commitment period, 2008 to 2012. In the reference scenarios, these policies were modeled as not continuing after 2012. In the stabilization scenarios, these initial period policies were superseded by the long-term control strategies imposed by each group.

Participation in Stabilization Scenarios

For the stabilization scenarios, it was assumed that policies to limit the change in radiative forcing would be applied globally after 2012, as directed by the Prospectus. Although it seems unlikely that all countries would simultaneously join such a global agreement, and the economic costs of stabilization would be greater with less-than-universal participation, the assumption that all countries participate does provide a useful benchmark.

Policy Instrument Assumptions in Stabilization Scenarios

Note that the issue of economic efficiency applies across both space and time. All of the scenarios assume an economically efficient allocation of reductions among nations in each time period, that is, across space. Thus, in these scenarios, GHG emissions in all regions and across all sectors of the economy were controlled by imposing a single price for each GHG at any point in time. As will be discussed in detail in Section 4.5, the prices of emissions for individual GHGs differ across the models. The implied ability to access emissions reduction opportunities wherever they are cheapest is sometimes referred to as *where* flexibility (Richels et al. 1996).

Timing of CO_2 Emissions Mitigation

The cost of stabilizing radiative forcing to any given level depends on the timing of the associated emissions reductions. There is a strong economic argument that costs will be lower if emissions reductions start slowly and then progressively ramp up, particularly for CO_2. Distributing emissions mitigation over time, such that larger efforts are undertaken later, reduces the current cost as a consequence of such effects as discounting, the preservation of energy-using capital stock over its natural lifetime, and the potential for the development of increasingly cost-effective technologies (Wigley et al. 1996).

Although 100 years is a very long time horizon for economic scenarios, it is not long enough to fully evaluate stabilization goals. For several of the radiative forcing stabilization levels, the scenarios are only approaching stabilization in 2100; radiative forcing is below the long-term stabilization levels and still rising, but the rate of increase is slowing. Stabilizing radiative forcing and associated atmospheric GHG concentrations requires that any emissions be completely offset by uptake or destruction processes. Because ocean and terrestrial uptake of CO_2 is subject to saturation and system inertia, at least for the approximate CO_2 concentration levels considered in this research, emissions need to peak and subsequently decline during the twenty-first century or soon thereafter. In the very long term (many hundreds to thousands of years), emissions must de-

cline to virtually zero for any CO_2 concentration to be maintained. Although there is some flexibility in the inter-temporal allocation of emissions, this allocation is inherently constrained by the carbon cycle. Given that anthropogenic CO_2 emissions rise with time in all three of the reference scenarios, the degree of CO_2 emissions reduction also increases steadily with time in the stabilization scenarios.

Different approaches were used by the modeling groups to determine the profile of emissions reductions over time and how the different GHGs contribute to meeting the radiative forcing stabilization levels. A major reason for the difference is the structure of the models. MERGE is an inter-temporal optimization model and is able to solve for the cost-minimizing allocation of emissions reductions across GHGs and over time to meet a given radiative forcing stabilization level. It thus offers insights regarding the optimal path of emissions reductions. A positive discount rate will lead to a gradual phase-in of emissions reductions, and the tradeoff among GHGs is endogenously calculated based on the contribution each makes toward the long-term goal (Manne and Richels 2001). The changing relative prices of GHGs over time can be interpreted as an optimal trading index for the GHGs that combines economic considerations with modeled physical considerations (lifetime and radiative forcing). The resulting relative weights are different from those derived using Global Warming Potential (GWP) indices, which are based purely on physical considerations (IPCC 2001). Furthermore, economically efficient indices for the relative importance of GHG emissions reductions will vary over time and across policy regimes.

IGSM and MiniCAM are simulation models and do not endogenously solve for optimal allocations over time and by GHG. However, the choice of price paths over time used in the stabilization scenarios for the IGSM and MiniCAM modeling groups take account of insights from economic principles that lead to a pattern similar to that computed by MERGE. The pattern was anticipated by Peck and Wan (1996) using a simple optimizing model with a carbon cycle and by Hotelling (1931) in a simpler context.

In the MiniCAM stabilization scenarios, the rate of increase in the carbon price was set equal to the rate of interest plus the average rate of carbon removal from the atmosphere by natural systems. This approach follows Peck and Wan (1996) and yields a resulting carbon price path similar in structure to that obtained in the MERGE scenarios. This carbon price path ensures that the present discounted marginal cost of having one tonne of carbon less in the atmosphere during one period in the future is exactly the same regardless of whether the removal takes place today or one period later. When marginal costs are equal over time, total costs cannot be reduced by making emissions mitigation either earlier or later.

As is the case in the MERGE scenarios, the exponential increase in the price of CO_2 continues until such time as radiative forcing is stabilized in the MiniCAM stabilization scenarios. Thereafter, the price is set by the carbon cycle. That is, once radiative forcing has risen to its stabilization level, additional CO_2 can only enter the atmosphere to the extent that natural processes remove it, otherwise CO_2 radiative forcing would be increasing. This is relevant in the Level 1 stabilization scenario and, to a lesser extent, in the Level 2 stabilization scenario. However, it is not relevant in the Level 3 or Level 4 scenarios because stabilization is not reached until after the end of the twenty-first century.

The IGSM scenarios are based on a carbon price path that rises 4% per year. The initial carbon price is set to achieve the required concentrations and radiative forcing. Thus, the rate of increase in the CO_2 price paths is identical for all stabilization scenarios, but the initial value of the carbon price is different. The lower the concentration of CO_2 allowed, the higher the initial price. The insight behind this approach is that an entity faced with a carbon constraint and a decision to reduce emissions now or later would compare the expected return on that emissions reduction investment with the rate of return elsewhere in the economy. The 4% rate is taken to be this economy-wide rate of return. If the carbon price were rising more rapidly than the rate of return, investments in emissions reductions would yield a higher return than investments elsewhere in the economy, so that the entity would invest more in emissions reductions now (and possibly bank emissions permits to use them later). By the same logic, an increase in the carbon price lower than the rate of return would lead to a decision to postpone emissions reductions. It would lead to a tighter carbon constraint and a higher carbon price in the future. Thus, this approach is intended to be consistent with a market solution that would allocate emissions reductions through time.

Timing of Non-CO_2 Emissions Mitigation

Like CO_2, the contribution of non-CO_2 GHGs to radiative forcing depends on their concentrations. However, these gases are dissociated in the atmosphere over time so that the relationship between emissions and concentrations is different from that for CO_2, as are the sources of emissions and opportunities for emissions reductions. Each of the three modeling groups used its own approach to model control of non-CO_2 GHGs. As noted above, MERGE employs an inter-temporal optimization approach. The price of each GHG was determined so as to minimize the cost of stabilizing radiative forcing at each level. Thus, the price of each GHG was constant across regions at any point in time, but varied over time so as to minimize the cost of achieving each stabilization level.

In the MiniCAM stabilization scenarios, non-CO_2 GHG prices were tied to the price of CO_2 using the GWPs of the gases. This procedure has been adopted by parties to the Kyoto Protocol and applied in the definition of the U.S. emissions intensity goal. The IGSM stabilization scenarios are based on the same approach as MiniCAM stabilization scenarios for determining the prices for HFCs, PFCs, and SF_6, pegging the prices to that of CO_2 using GWP coefficients. For CH_4 and N_2O, however, independent emission stabilization levels were set for each gas in the IGSM scenarios because GWPs poorly represent the full effects of CH_4, and emissions trading at GWP rates leads to problems in defining what stabilization means when CH_4 and N_2O are involved (Sarofim et al. 2005). The relatively near-term stabilization for CH_4 in the IGSM scenarios implies that near-term emissions reductions result in economic benefit, an approach consistent with a view that there are risks associated with levels of radia-

tive forcing below the long-term stabilization levels. This approach is different than that followed in the MERGE scenarios, where any value of CH_4 emissions reductions is derived only from the extent to which it contributes to meeting the long-term stabilization level. In the MERGE stabilization scenarios, reductions of emissions of short-lived species like CH_4 have very little consequence for a radiative forcing stabilization level that will not be reached for many decades, so the optimized result places little value on reducing emissions of short-lived species until the stabilization level is approached. A full analysis of the resulting climate change and its effects would be required to select between the approaches used in the MERGE and IGSM scenarios. The different stabilization paths in the scenarios from these two models provide a range of plausible scenarios for non-CO_2 GHG stabilization. The MiniCAM scenarios yield an intermediate result.

IMPLICATIONS FOR RADIATIVE FORCING, GREENHOUSE GAS CONCENTRATIONS, AND EMISSIONS

Despite significantly different radiative forcing levels in the reference scenarios, radiative forcing relative to preindustrial levels in 2100 is similar across models in all four stabilization scenarios. CO_2 concentrations are also similar in 2100 across the models. Scenarios with higher CO_2 concentrations for a given stabilization level generally have lower concentrations and emissions of non-CO_2 GHGs, trading off reductions in these substances to make up for higher forcing from CO_2.

All three modeling groups produced scenarios in which emissions reductions below levels in the reference scenarios were much smaller between 2000 and 2050 than between 2050 and 2100. With one exception at the least stringent stabilization level, the stabilization scenarios were characterized by a peak and decline in global CO_2 emissions in the twenty-first century. In the most stringent scenarios, CO_2 emissions begin to decline immediately or within a matter of decades.

		Radiative Forcing in 2100 (W/m^{-2} from preindustrial)		
Stabilization Level	Long-Term Radiative Forcing Limit (W/m^2 from preindustrial)	IGSM	MERGE	MiniCAM
Reference	No Constraint	8.6	6.6	6.4
Level 4	6.7	6.1	6.2	6.1
Level 3	5.8	5.4	5.7	5.5
Level 2	4.7	4.4	4.7	4.5
Level 1	3.4	3.5	3.4	3.4

Table 4.1. Radiative Forcing in the Year 2100 Across Scenarios

Implications for Radiative Forcing

Given that all the models were constrained to the same radiative forcing stabilization levels, radiative forcing from preindustrial for the year 2100 is similar across the models (Table 4.1).[1] The differences across the models between the long-term stabilization levels and the radiative forcing levels in 2100 are smaller for Levels 1 and 2 than for Levels 3 and 4 because the latter allow a greater accumulation of GHGs in the atmosphere. For Levels 3 and 4, each modeling group required radiative forcing to be below the long-term limits in 2100 to allow for subsequent emissions to fall gradually toward levels required for stabilization.

The radiative forcing stabilization paths are shown in Figure 4.1. Even though they reflect different criteria used to allocate emissions reductions over time, the paths are very similar across models. The radiative forcing paths are dominated by radiative forcing associated with CO_2 concentrations, which in turn are driven by cumulative emissions. Thus, even fairly different time profiles of CO_2 emissions can yield relatively little difference in concentrations and radiative forcing.

Although their totals are similar, the GHG composition of radiative forcing differs among the models. Figure 4.2 shows the breakdown among gases in 2100 for the reference scenario along with all four stabilization levels. Forcing is dominated by CO_2 in all scenarios at all stabilization levels, but there are variations among

models. For example, the MiniCAM stabilization scenarios have larger contributions from CO_2 and lower contributions from the non-CO_2 gases than the scenarios from the other two models. Conversely, the MERGE stabilization scenarios have higher contributions from the non-CO_2 gases and lower contributions from CO_2 relative to the IGSM and MiniCAM stabilization scenarios.

Implications for Greenhouse Gas Concentrations

The relative GHG composition of radiative forcing across models in any scenario reflects differences in concentrations of the GHGs. The CO_2 concentration paths are presented in Figure 4.3, and the year 2100 atmospheric levels are shown in Table 4.3. Because the stabilization levels were specified in terms of total radiative forcing from the multiple GHGs, it is possible to meet those levels while varying from the approximate CO_2 concentration levels used to construct them (Table 1.2). That means CO_2 concentrations in 2100 differ across models for any stabilization level. For example, the CO_2 concentrations in the MiniCAM stabilization scenarios are generally higher than in IGSM and MERGE stabilization scenarios. Consequently, CH_4 and N_2O concentrations are systematically lower as can be seen in Figure 4.4 and Figure 4.5.

Differences in the GHG concentrations among the scenarios from the three models reflect differences in the way that tradeoffs were made among gases and differences in assumed emissions reduction opportunities for non-CO_2 GHGs compared to CO_2.

---

[1] The IGSM exceeds the Level 1 target by 0.1 W/m^2, which is a negligible difference that results from the iterative process required to achieve a radiative forcing target.

Figure 4.1. Total Radiative Forcing by Year Across Scenarios (W/m² from preindustrial). Radiative forcing trajectories differ across the stabilization levels but are similar among models for each stabilization level. The similarity across models reflects the design of the scenarios. Radiative forcing is stabilized or close to being stabilized this century in the Level 1 and Level 2 scenarios. Radiative forcing remains below the long-term radiative forcing stabilization level in 2100 in the Level 3 and Level 4 stabilization scenarios, allowing for a gradual approach to stabilization in the following century.

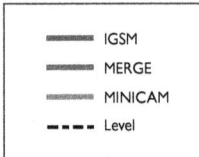

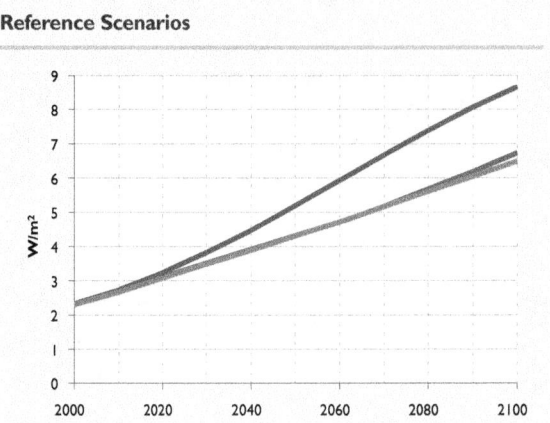

Reference Scenarios

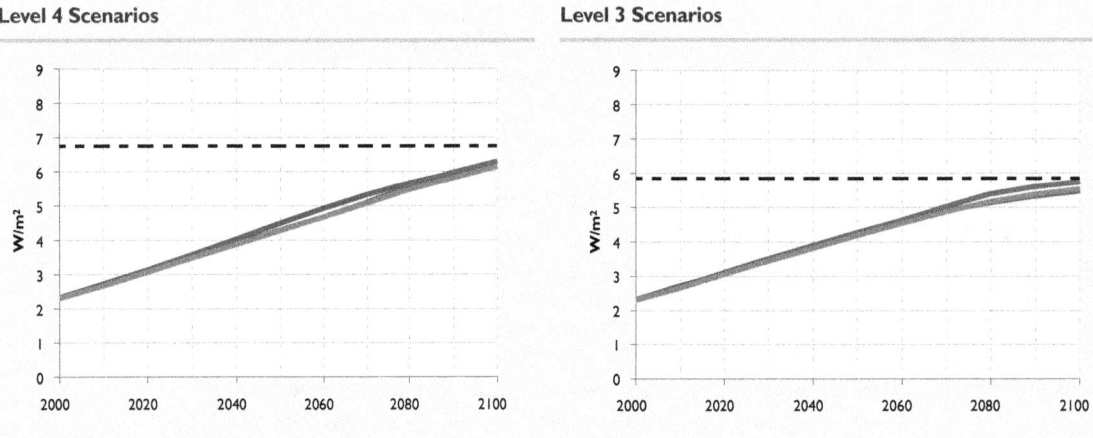

Level 4 Scenarios

Level 3 Scenarios

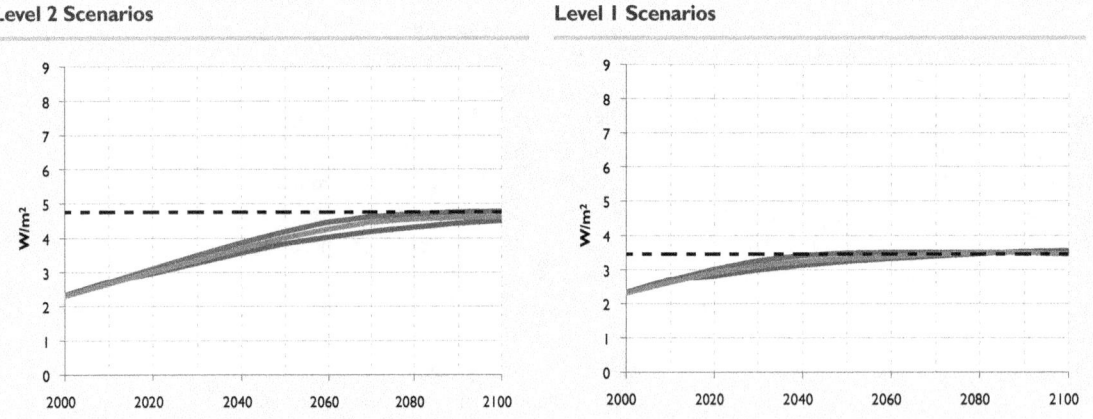

Level 2 Scenarios

Level 1 Scenarios

Approximate stabilization of CO_2 concentrations occurs by 2100 in all the Level 1 and Level 2 scenarios, but concentrations are still increasing in 2100 for the Level 3 and Level 4 scenarios, although at a slowing rate. An important implication of the less stringent stabilization levels is that substantial emissions reductions would be required after 2100. Sometime within the next century, all the stabilization paths would require emissions levels nearly as low as that for Level 1. Higher stabilization levels do not change the nature of long-term changes in emissions required in the global economy; they only delay when the emissions reductions must be achieved.

Figure 4.2. Total Radiative Forcing by Gas in 2100 Across Scenarios (W/m² from preindustrial). CO₂ is the main contributor to radiative forcing by the end of the century in the scenarios from all three modeling groups. The IGSM reference scenario has the highest contribution from non-CO₂ GHGs among the three models. The MERGE stabilization scenarios have the highest contribution from non-CO₂ GHGs among the three models, implying greater non-CO₂ control efforts in the IGSM scenarios than in the MERGE scenarios. Contributions from non-CO₂ GHGs are lowest in the MiniCAM scenarios, reflecting, in part, assumptions about control of these substances for non-climate reasons.

Reference Scenarios

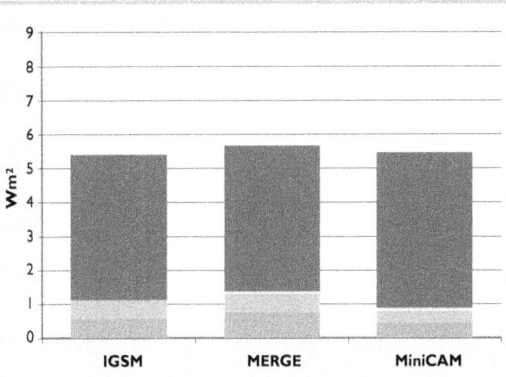

Level 4 Scenarios

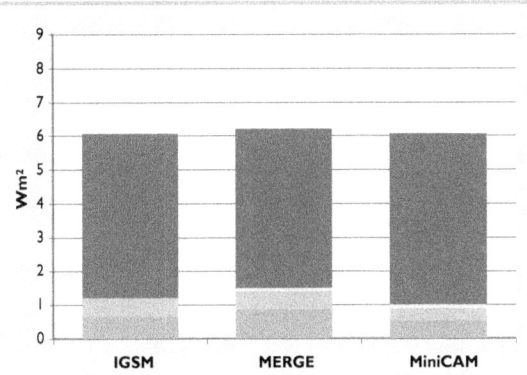

Level 3 Scenarios

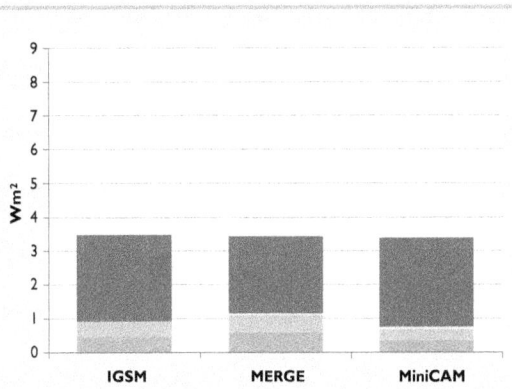

Level 2 Scenarios

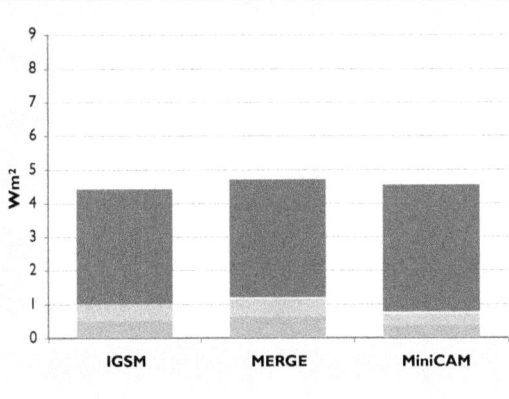

Level 1 Scenarios

In all the stabilization scenarios, as the rise in atmospheric concentrations slows, ocean uptake slows and even begins to decline. These natural removal processes are uncertain, and to some extent this uncertainty is reflected in differences in the scenarios from the three modeling groups, as shown in Figure 4.6. Ocean uptake is smallest in the IGSM scenarios. The MERGE scenarios have the highest uptake for the least stringent stabilization levels, and the MiniCAM and MERGE scenarios are almost identical under the most stringent stabilization levels.

Figure 4.3. CO$_2$ Concentrations Across Scenarios (ppmv). Atmospheric concentrations of CO$_2$ range from about 700 ppmv to 900 ppmv in 2100 in the reference scenarios, with no sign of slowing. In the stabilization scenarios, differences in CO$_2$ concentrations among models occur because of the relative contribution of other GHGs to meeting the radiative forcing stabilization levels, and because for Levels 3 and 4, the scenarios are based on a gradual approach to the stabilization level that will not be reached until the following century.

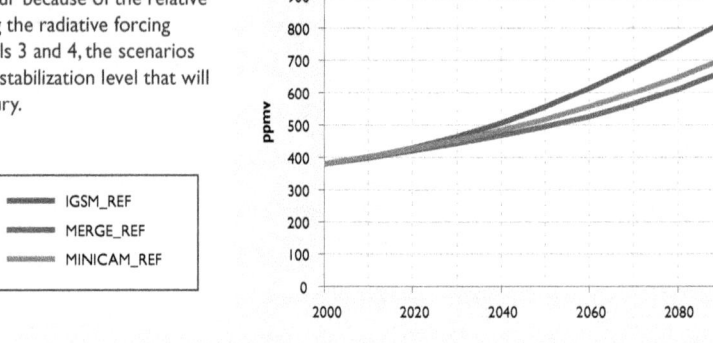

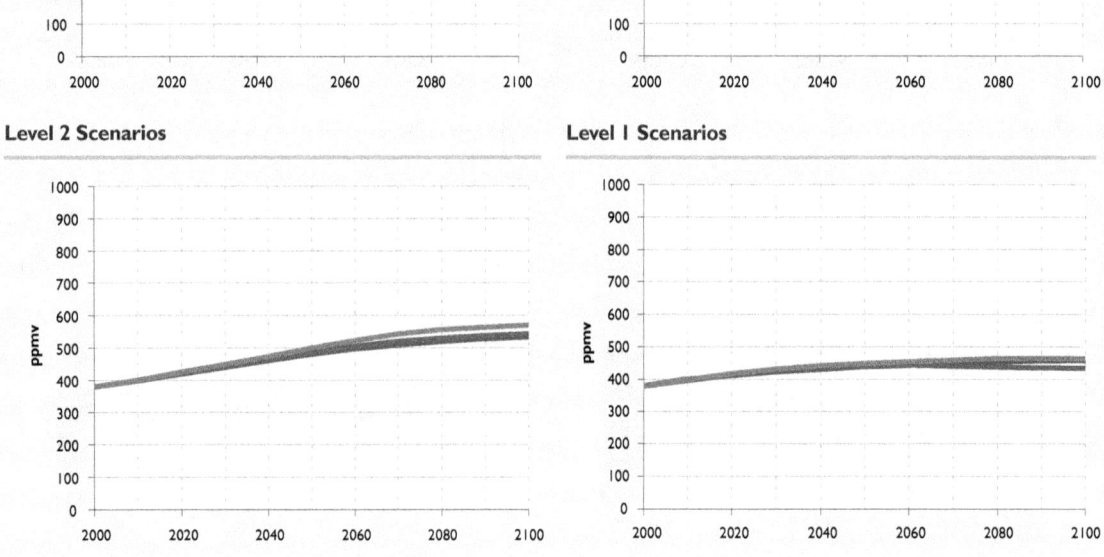

Implications for Greenhouse Gas Emissions

IMPLICATIONS FOR GLOBAL CO$_2$ EMISSIONS
Global CO$_2$ emissions begin declining immediately after 2010 or in a matter of decades in all three Level 1 stabilization scenarios (Figure 4.7). The constraint is so tight that there is relatively little room for variation among models.

All three modeling groups show continued emissions growth throughout the first half of the twenty-first century for Level 4, the least stringent stabilization levels, and the MiniCAM

	CO$_2$ Concentration in 2100 (ppmv)			
Level	Approximate Long-Term CO$_2$ Concentration Limit (ppmv)	IGSM	MERGE	MiniCAM
Reference	—	875	711	746
Level 4	750	677	670	716
Level 3	650	614	619	656
Level 2	550	526	535	562
Level 1	450	451	426	456

Table 4.2. CO$_2$ Concentrations in the Year 2100 Across Scenarios (ppmv). The approximate CO$_2$ concentrations were used as a guide to develop the radiative forcing stabilization levels. The scenarios were required to meet the total radiative forcing limits. The CO$_2$ concentrations in the scenarios do not exactly match these approximations and differ among the modeling groups because of differences in the treatment of the forces that influence emissions of GHGs, possibilities for emissions reductions, and tradeoffs between reductions among GHGs.

Figure 4.4. CH$_4$ Concentrations Across Scenarios (ppbv). Differences among the models in CH$_4$ concentrations are larger than differences in CO$_2$ concentrations. These differences stem from differences in reference scenarios, assumptions about options for emissions reductions, and the methods used by the modeling groups for determining the relative emissions reductions among different GHGs. Reductions in non-CO$_2$ GHG emissions in the MiniCAM stabilization scenarios are based on 100-year GWPs. The MERGE stabilization scenarios are based on intertemporal optimization, leading to relatively little value for controlling CH$_4$ emissions until the stabilization level is approached due to the relatively short lifetime of CH$_4$. The IGSM stabilization scenarios are based on independent stabilization of CH$_4$.

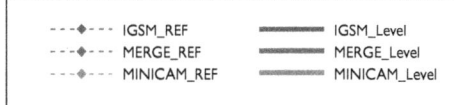

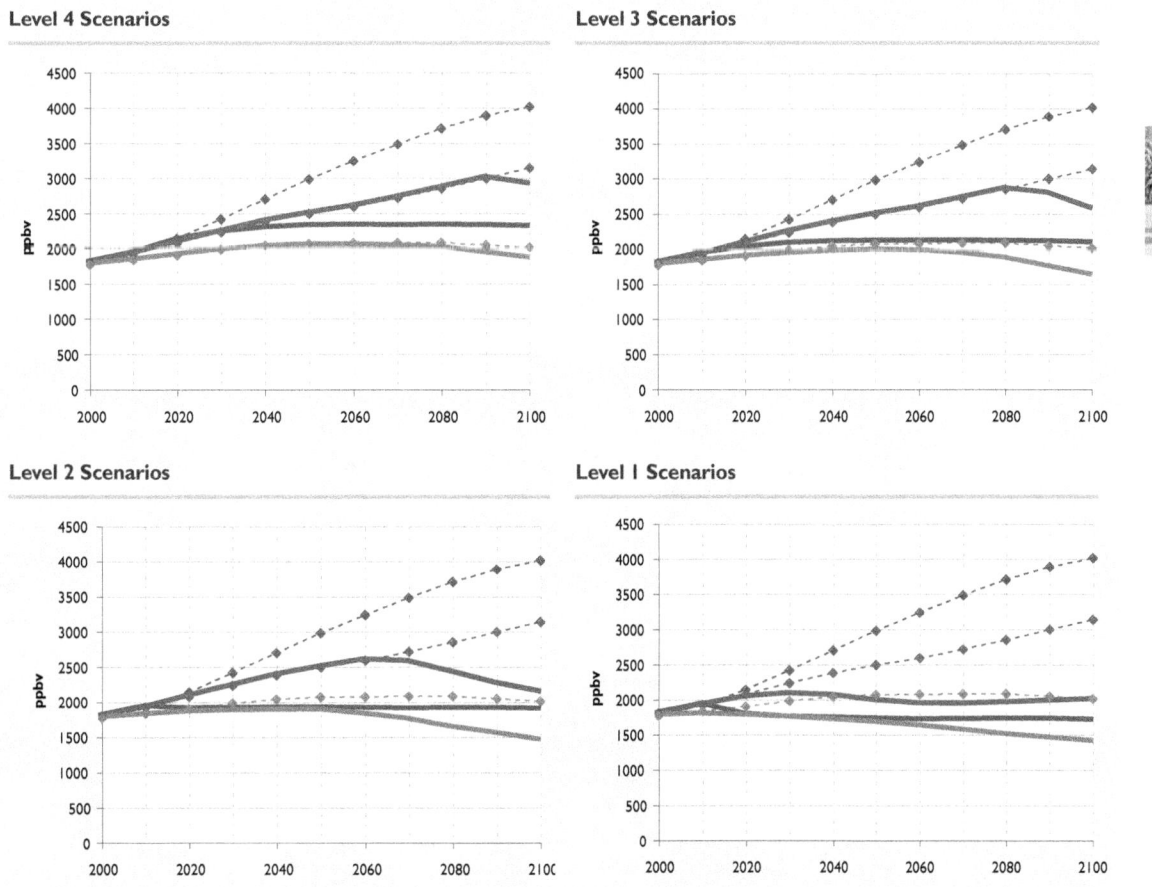

Figure 4.5. N₂O Concentrations Across Scenarios (ppbv). Atmospheric concentrations of N₂O range from about 375 ppbv to 500 ppbv in 2100 across the scenarios, with concentrations continuing to rise in the reference scenarios. Different approaches were used by the different modeling groups to develop emissions reductions, leading to differences in concentrations between the reference and stabilization scenarios.

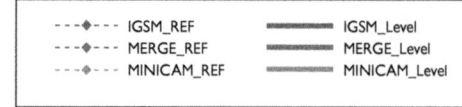

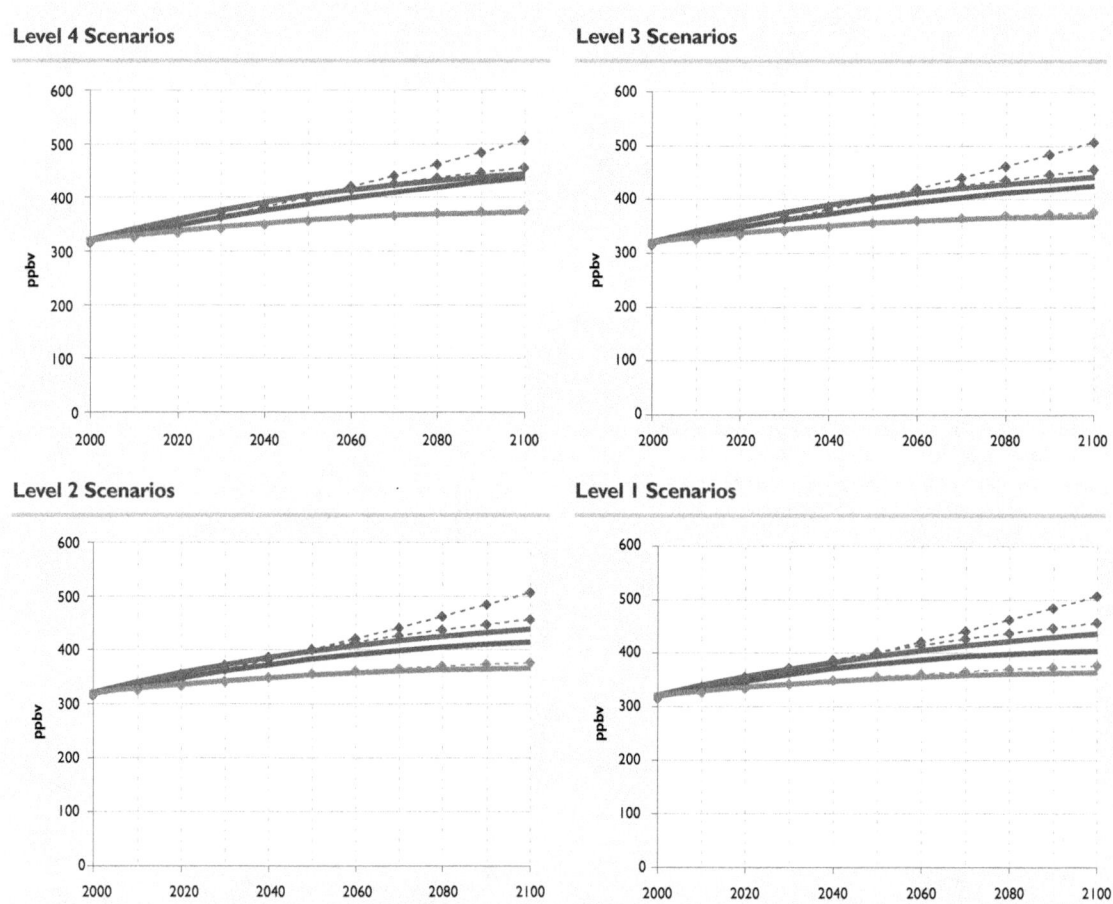

Level 4 scenario exhibits increasing emissions throughout the century, although emissions are approaching a peak by 2100.

The stabilization scenarios from all three modeling groups exhibit more emissions reduction in the second half of the twenty-first century than in the first half, as noted earlier, so the mitigation challenge grows with time. The precise timing and degree of departure from the reference scenario depend on many aspects of the scenarios and on each model's representation of Earth system properties, including the radiative forcing stabilization level, the carbon cycle, atmospheric chemistry, the character of technology options over time, the reference scenario CO₂ emissions path, the non-climate policy environment, the rate of discount, and the climate

policy environment. For Level 4, 85% or more of emissions mitigation occurs in the second half of the twenty-first century in the scenarios from all three modeling groups. Even for Level 1, where the limit on radiative forcing is the tightest and near-term mitigation most urgent, 75% or more of the emissions reduction below reference scenario occurs in the second half of the century. While this is partly a result of the *when* flexibility assumption, continuing emissions growth in the reference scenarios means that the percentage reduction increases over time.

All three of the modeling groups constructed reference scenarios in which Non-Annex 1 emissions were a larger fraction of the global total in the future than at present (Figure 3.16).

Figure 4.6. Ocean CO$_2$ Uptake Across Scenarios (GtC/yr, expressed in terms of net emissions). Oceans have taken up approximately one half of anthropogenic emissions of CO$_2$ since preindustrial times, and future ocean behavior is an important determinant of atmospheric concentrations. The three-dimensional ocean used for the IGSM scenarios shows the least ocean carbon uptake and considerable slowing of carbon uptake even in the reference scenario as carbon concentrations continue to rise. The MERGE reference scenario shows the largest uptake among the three models, and the MERGE stabilization scenarios have the greatest reductions from the reference scenario among the models. The MiniCAM scenarios are intermediate at most stabilization levels. At the more stringent stabilization levels, the MERGE and MiniCAM scenarios exhibit similar ocean uptake behavior.

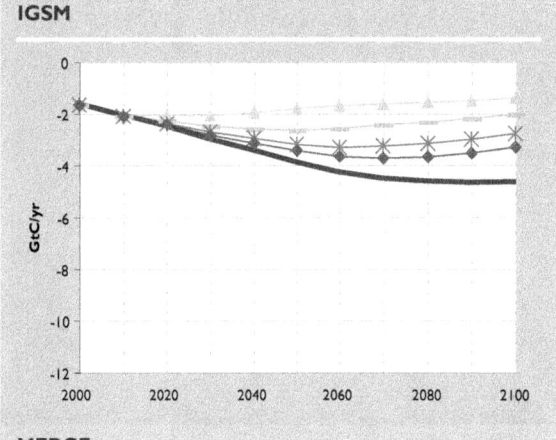

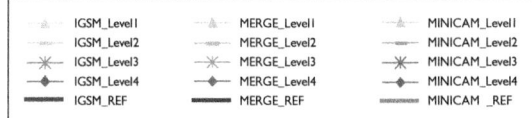

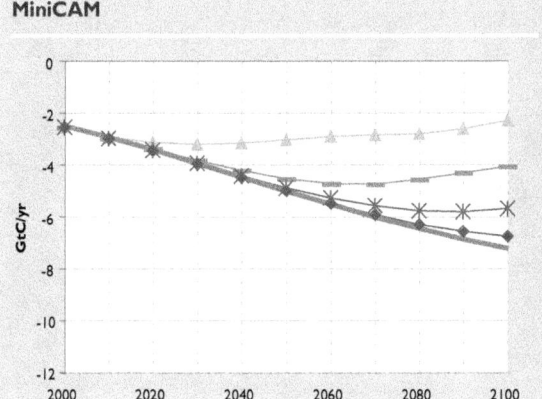

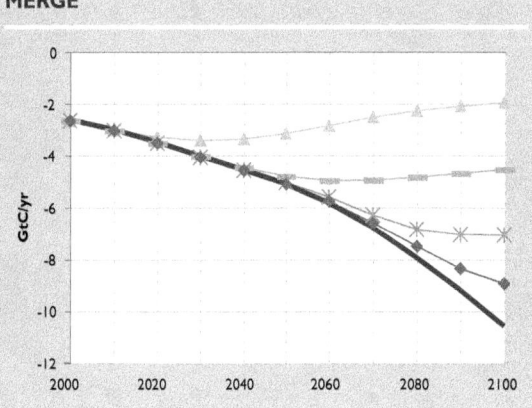

Because the stabilization scenarios are based on the assumption that all regions of the world face the same price of GHG emissions and have access to the same general set of technologies for mitigation, the resulting distribution of emissions mitigation between Annex I and Non-Annex I regions generally reflects the distribution of reference scenario emissions among them. So, when radiative forcing is restricted to Level 1, all three models find that more than half of the emissions mitigation occurs in Non-Annex I regions by 2050 because more than half of reference scenario emissions occur in Non-Annex I regions. Note that with the global policy specified so that a common carbon price occurs in all regions at any one time, emissions reductions occur separately from and mostly independent of the distribution of the economic burdens of reduction.

IMPLICATIONS FOR NON-CO$_2$ GREENHOUSE GAS EMISSIONS

The stabilization properties of the non-CO$_2$ GHGs differ due to their lifetimes (as determined by chemical reactions in the atmosphere), technologies for reducing emissions, and natural sources. CH$_4$ has a relatively short lifetime, and anthropogenic sources are a big part of CH$_4$ emissions. If anthropogenic emissions are kept constant, an approximate equilibrium between oxidation and net emissions will be established relatively quickly and concentrations will stabilize. The same is true for the relatively short-lived HFCs.

CH$_4$ emissions under stabilization are systematically lower the more stringent the stabilization level, as can be seen in Figure 4.8. The MiniCAM scenarios have the lowest CH$_4$ emis-

Figure 4.7. Fossil Fuel and Industrial CO$_2$ Emissions Across Scenarios (GtC/yr). Fossil fuel CO$_2$ emissions vary among the reference scenarios, but the three differing emissions trajectories lead to emissions in 2100 in the range of 22.5 GtC/yr to 24.0 GtC/yr. The timing of emissions reductions varies substantially across the stabilization levels. In the Level 1 scenarios, global emissions begin to decline soon after the stabilization policy is put in place (as the scenarios were designed, after 2012), and emissions are below current levels by 2100 in all of the Level 1 and Level 2 scenarios. Emissions peak sometime around the mid-century to early in the next century in the Level 3 and Level 4 scenarios and then begin a decline that would continue beyond 2100.

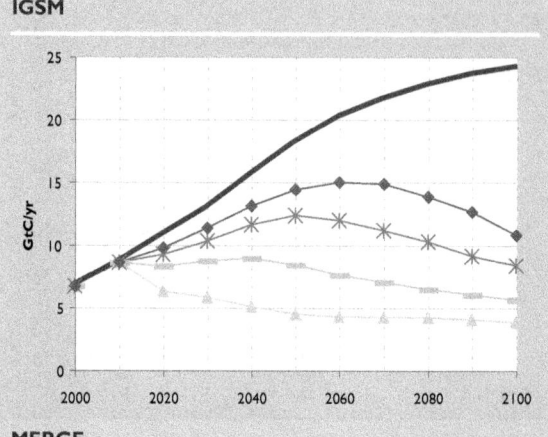

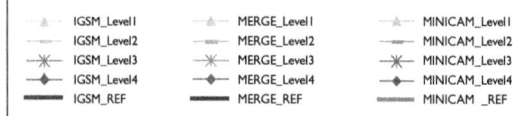

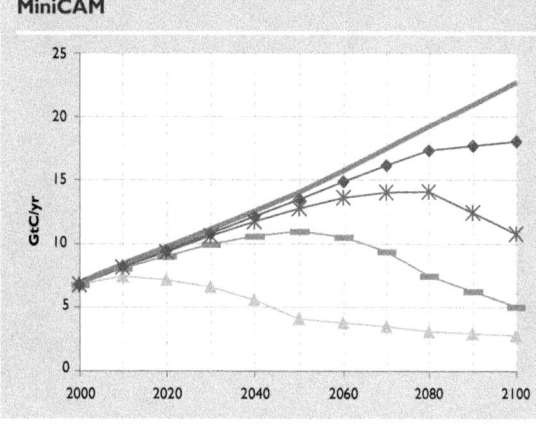

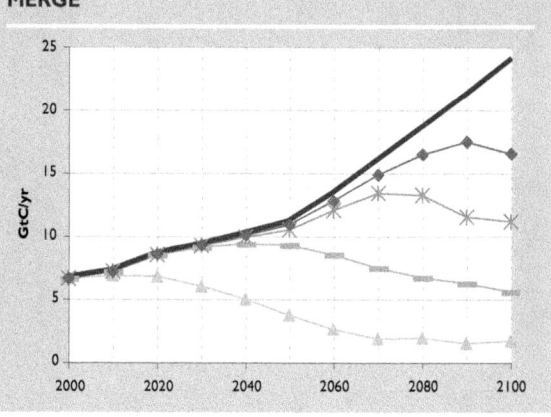

sions among the models in the reference scenario and the stabilization scenarios. The assumed policy environment for CH$_4$ control is also important. Despite the fact that the IGSM reference scenario has higher reference CH$_4$ emissions than the MERGE reference scenario, the MERGE stabilization scenarios have higher emissions under stabilization in several instances. The reason is that the MERGE intertemporal optimization approach leads to a low relative price for CH$_4$ emissions in the near term, which grows rapidly relative to CO$_2$, favoring strong reductions of CH$_4$ emissions only toward the end of the century, whereas CH$_4$ emissions were controlled based on quantitative limits in the IGSM stabilization scenarios, and these limits lead to substantial reduction early in the century. Thus, emissions in the MERGE stabilization scenarios sometimes exceed those in the IGSM stabilization scenarios until the relative CH$_4$ price rises sufficiently to induce substantial emissions reductions.

The very long-lived gases are nearly indestructible, thus for stabilization their emissions must be very near zero. Based on the assumptions used by all three modeling groups, it is possible, at reasonable cost, to achieve substantial reductions in long-lived gas emissions. While these substances are important, their emissions are not as difficult to reduce as those from fossil energy.

N$_2$O is more problematic. A major anthropogenic source is from use of fertilizer for agricultural crops – an essential use. Moreover, its natural sources are important, and they are augmented by terrestrial changes associated with climate change. It is fortunate that N$_2$O is not a major contributor to radiative forcing because the technologies and strategies needed to achieve its stabilization are not obvious at this time. Nevertheless, differences in the control of N$_2$O are observed across models, as shown in Figure 4.9, although these differences are smaller than those for CH$_4$.

Figure 4.8. CH₄ Emissions Across Scenarios (Mt CH₄/yr). Emissions of anthropogenic CH₄ vary widely across the models, including differences in year 2000 emissions that reflect uncertainty about these emissions. With current concentrations and destruction rates relatively well known, the difference in current levels means that IGSM scenarios ascribe relatively more to anthropogenic sources and relatively less to natural sources than do the MERGE and MiniCAM scenarios. Wide differences in scenarios for the future reflect differing modeling approaches, outlooks for activity levels that lead to emission reductions, and assumptions about whether emissions will be reduced for non-climate reasons.

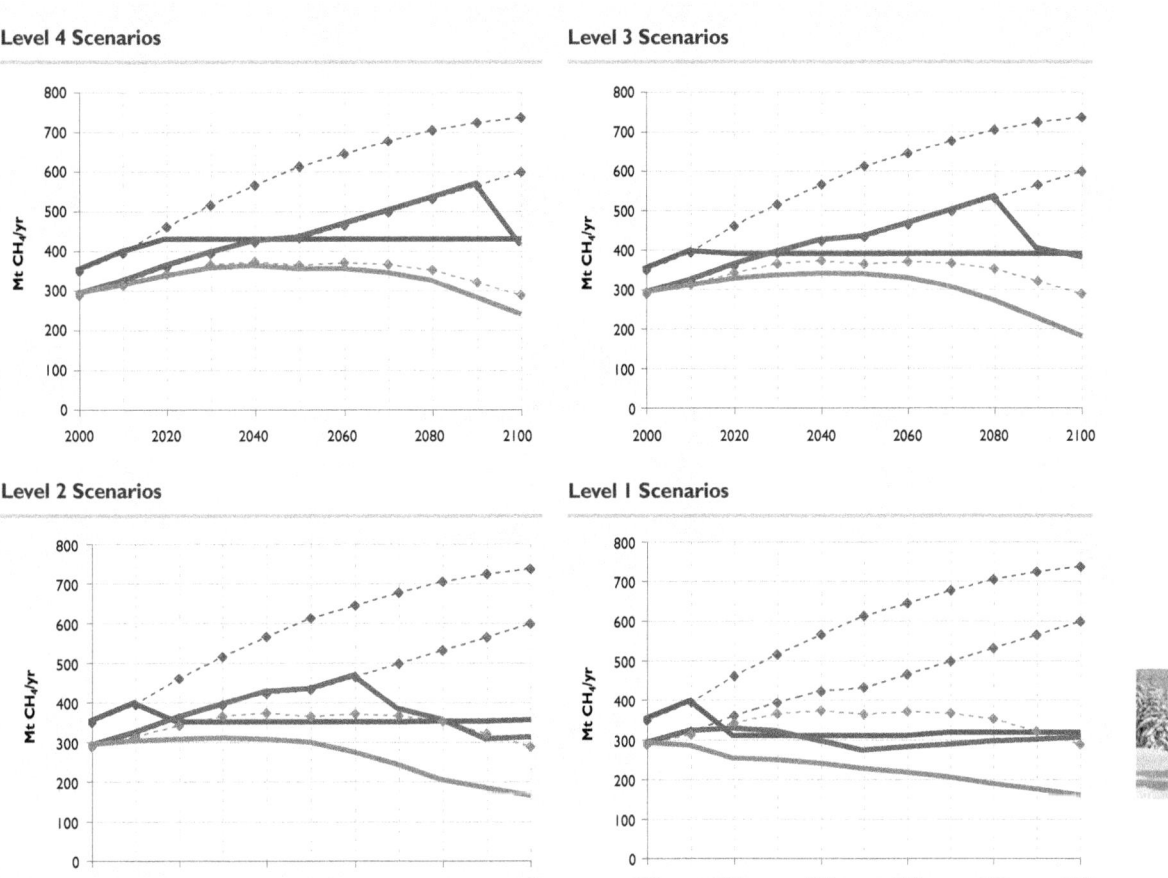

IMPLICATIONS FOR ENERGY USE, INDUSTRY, AND TECHNOLOGY

In these scenarios, stabilizing radiative forcing requires a transformation of the global energy system, including reductions in the demand for energy and changes in the mix of energy technologies and fuels. This transformation is more substantial and takes place more quickly at the more stringent stabilization levels. Fossil fuel use and energy consumption are reduced in all the stabilization scenarios due to increased consumer prices for fossil fuels. CO_2 emissions from electric-ity production are reduced at relatively lower prices than CO_2 emissions from other sectors, such as transport, industry, and buildings. Emissions are reduced from electric power by increased use of technologies such as CCS, nuclear energy, and renewable energy. Other sectors respond to rising greenhouse gas prices by reducing demands for fossil fuels; substituting low- or non-emitting energy sources such as bioenergy and low-carbon electricity or hydrogen; and applying CCS where possible.

Figure 4.9. N₂O Emissions Across Scenarios (Mt N₂O/yr).
Anthropogenic emissions of N₂O are similar across models in the stabilization scenarios despite large differences in the reference scenarios.

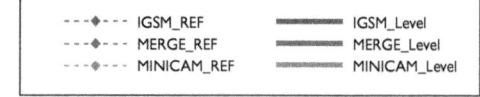

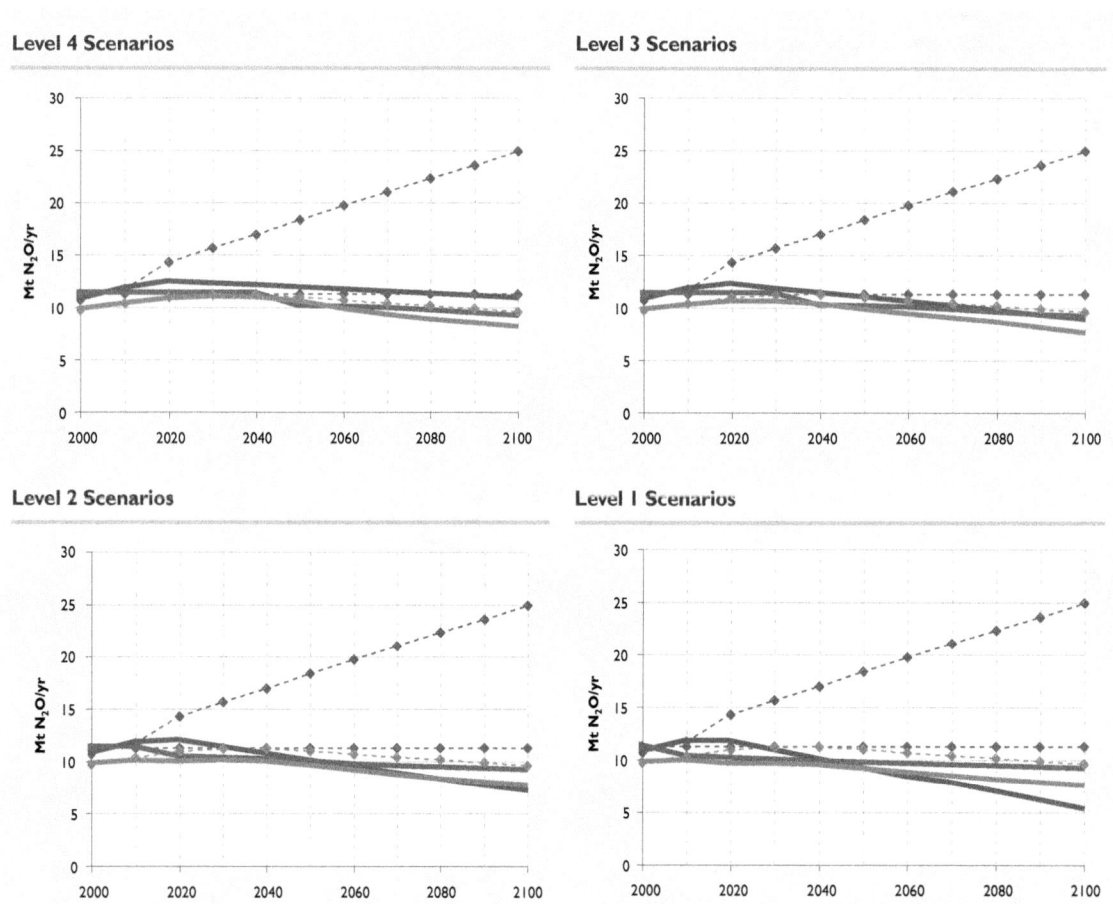

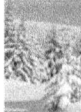

Changes in Global Energy Use

The degree and timing of change in the global energy system depends on the level at which radiative forcing is stabilized. Although differences in the reference scenarios developed by each of the three modeling groups led to different patterns of response, some important similarities emerge. The more stringent the radiative forcing stabilization level, the larger the change in the global energy system relative to the reference scenario; moreover, the scale of this change is increasing over time. Also, significant fossil fuel use continues at all four stabilization levels. This pattern can be seen in Figure 4.10, which shows the global primary energy consumption across the scenarios, and Figure 4.11, which shows the reference scenario from Chap-

ter 3 with an additional plot of the net changes in the various sources of primary energy for each stabilization level.

Although atmospheric stabilization would take away much of the growth potential of coal over the century, its usage expands above today's levels by the end of the century in all the stabilization scenarios. In several of the Level 1 and Level 2 scenarios, the global coal industry declines in the first half of the century before recovering by 2100 to levels of production somewhat larger than today. Oil and natural gas also continue as contributors to total energy over the century although, as with coal, they are increasingly pushed from the energy mix as the stabilization level is tightened.

One reason that fossil fuels continue to be utilized despite constraints on GHG emissions is that CCS technologies are available in the scenarios from all three modeling groups. Figure 4.10 shows that as the carbon price rises, CCS technology takes on an increasing market share. Section 4.4.2 addresses this pattern as well as the contribution of non-biomass renewable energy forms in greater detail.

Changes in the global energy system in response to constraints on radiative forcing reflect an interplay between technology options and the other assumptions that shaped the reference scenarios. For example, the MERGE reference scenario assumes relatively limited ability to access unconventional oil and gas resources and the evolution of a system that increasingly employs coal as a feedstock for the production of liquids, gases, and electricity. Against this background, a constraint on radiative forcing leads to reductions in coal use and end-use energy consumption. As the carbon price rises, nuclear and non-biomass renewable energy forms and CCS augment the response.

The IGSM scenarios assume greater availability of unconventional oil than the MERGE scenarios. Thus, the IGSM stabilization scenarios, in general, involve less reduction in coal use by the end of the century, but a larger decline in oil than in the MERGE stabilization scenarios. To produce liquid fuels for the transportation sector, the IGSM scenarios respond to a constraint on radiative forcing by growing biomass energy crops both earlier and more extensively than in the reference scenario. Also, reductions in energy demand are larger in the IGSM stabilization scenarios than in the scenarios from the other two models.

The MiniCAM stabilization scenarios include the smallest reductions in energy consumption among the models. The imposition of constraints on radiative forcing leads to reductions in oil, gas, and coal, as is the case with the IGSM and MERGE stabilization scenarios, but also leads to considerable expansion of nuclear power and renewable energy supplies. The largest supply response is in commercial bio-derived fuels. These fuels are largely limited to bio-waste recycling in the MiniCAM reference scenario. As the price of CO_2 rises, commercial

bioenergy becomes increasingly attractive. As will be discussed in Section 4.5, the expansion of the commercial biomass industry to produce hundreds of EJ/yr of energy has implications for crop prices, land use, land-use emissions, and unmanaged ecosystems.

The relative role of nuclear energy differs among the scenarios from the three modeling groups. The MERGE reference scenario deploys the largest amount of nuclear power, contributing 170 EJ/yr of primary energy in the year 2100. In the Level 1 stabilization scenario, deployment expands to 240 EJ/yr of primary energy in 2100. Nuclear power in the MiniCAM reference scenario produces 90 EJ/yr in the year 2100, which in the Level 1 stabilization scenario expands to more than 180 EJ/yr of primary energy in the year 2100. The IGSM stabilization scenarios show little change in nuclear power generation among the stabilization scenarios or compared with the reference, reflecting the assumption that nuclear levels are limited by policy decisions regarding safety, waste, and proliferation that are unaffected by climate policy.

Reductions in total primary energy consumption play an important role in all of the stabilization scenarios. In the IGSM stabilization scenarios, this is the largest single change in the global energy system. While not as dramatic as the IGSM stabilization scenarios, the MERGE and MiniCAM stabilization scenarios also exhibit reductions in energy demand. As will be discussed in Section 4.6, differences in primary energy reductions among the models reflect differences in the carbon prices required for stabilization, which are substantially higher in the IGSM stabilization scenarios than in the MERGE and MiniCAM stabilization scenarios. In all the stabilization scenarios, carbon price differences are reflected in the user prices of energy. Carbon prices, in turn, reflect technological assumptions that influence both the supply of alternative energy and the responsiveness of users to changing prices. The fuel and GHG prices discussed later in this chapter, therefore, can be instructive in understanding the character of technology assumptions employed in the models.

As noted throughout the preceding and following discussions, the economic equilibrium na-

Figure 4.10. Global Primary Energy Consumption by Fuel Across Scenarios (EJ/yr). The transition to stabilization, reflected most fully in the Level 1 stabilization scenarios, means an eventual phase-out of fossil fuel use unless CCS is employed. Consumption of non-fossil energy sources increases 6-fold to 14-fold over the century in the Level 1 stabilization scenarios. In the IGSM stabilization scenarios, more of the emissions reductions are met through demand reductions than in the scenarios from the other two modeling groups, with 2100 energy use cut by up to one-half relative to the reference scenario in 2100. In the MiniCAM Level 1 scenario, in contrast, total energy is reduced by less than 20%. Levels 2, 3, and 4 require progressively less transformation compared with the reference scenarios in the coming century, delaying these changes until beyond 2100. [Notes: i. Oil consumption includes that derived from tar sands and oil shales, and coal consumption includes that used to produce synthetic liquid and gaseous fuels. ii. Primary energy consumption from nuclear power and non-biomass renewable electricity are accounted for at the average efficiency of fossil-fired electric facilities, which vary over time and across scenarios. This long-standing convention means that, all other things being equal, increasing efficiency of fossil-electric energy lowers the contribution to primary energy from these sources.]

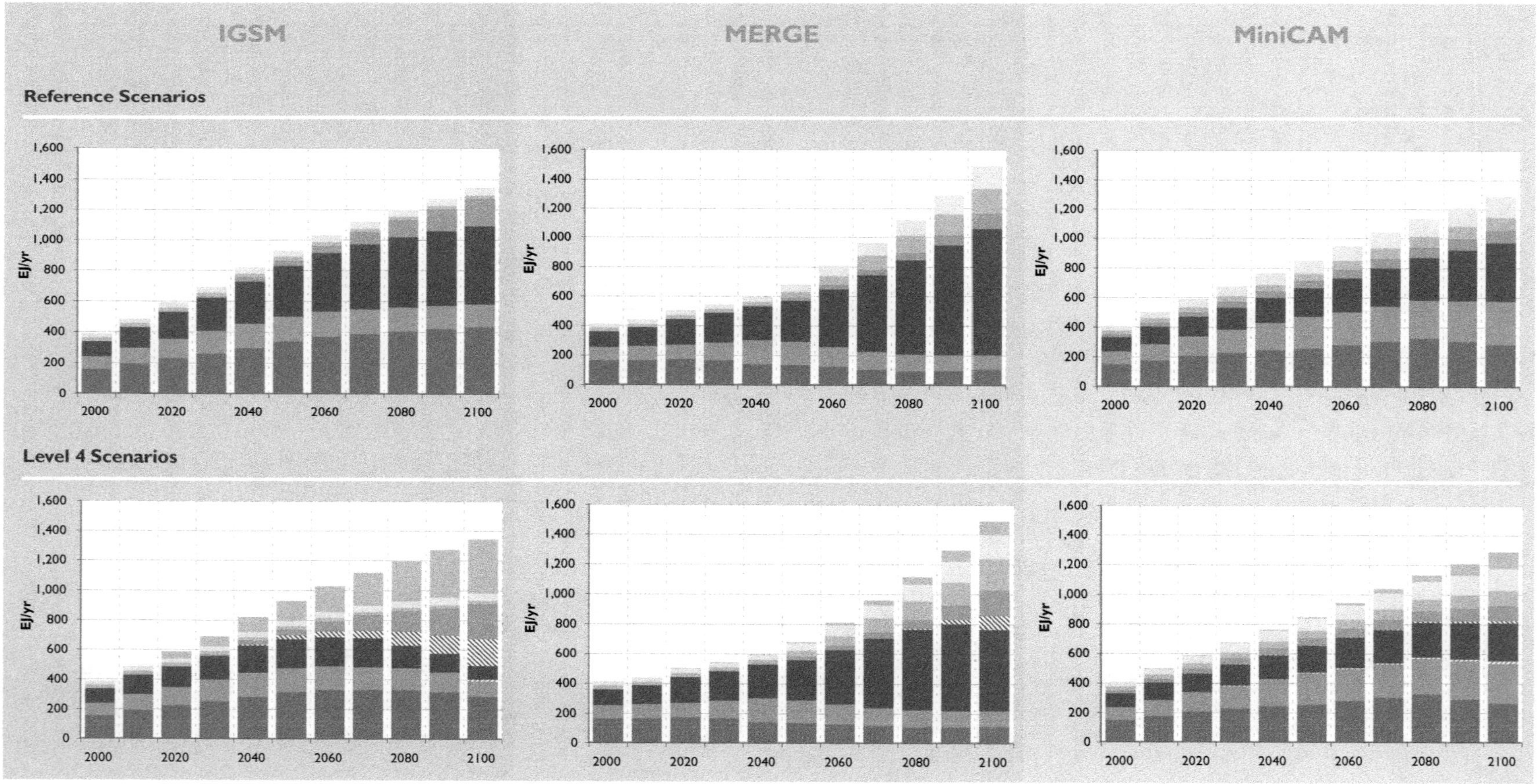

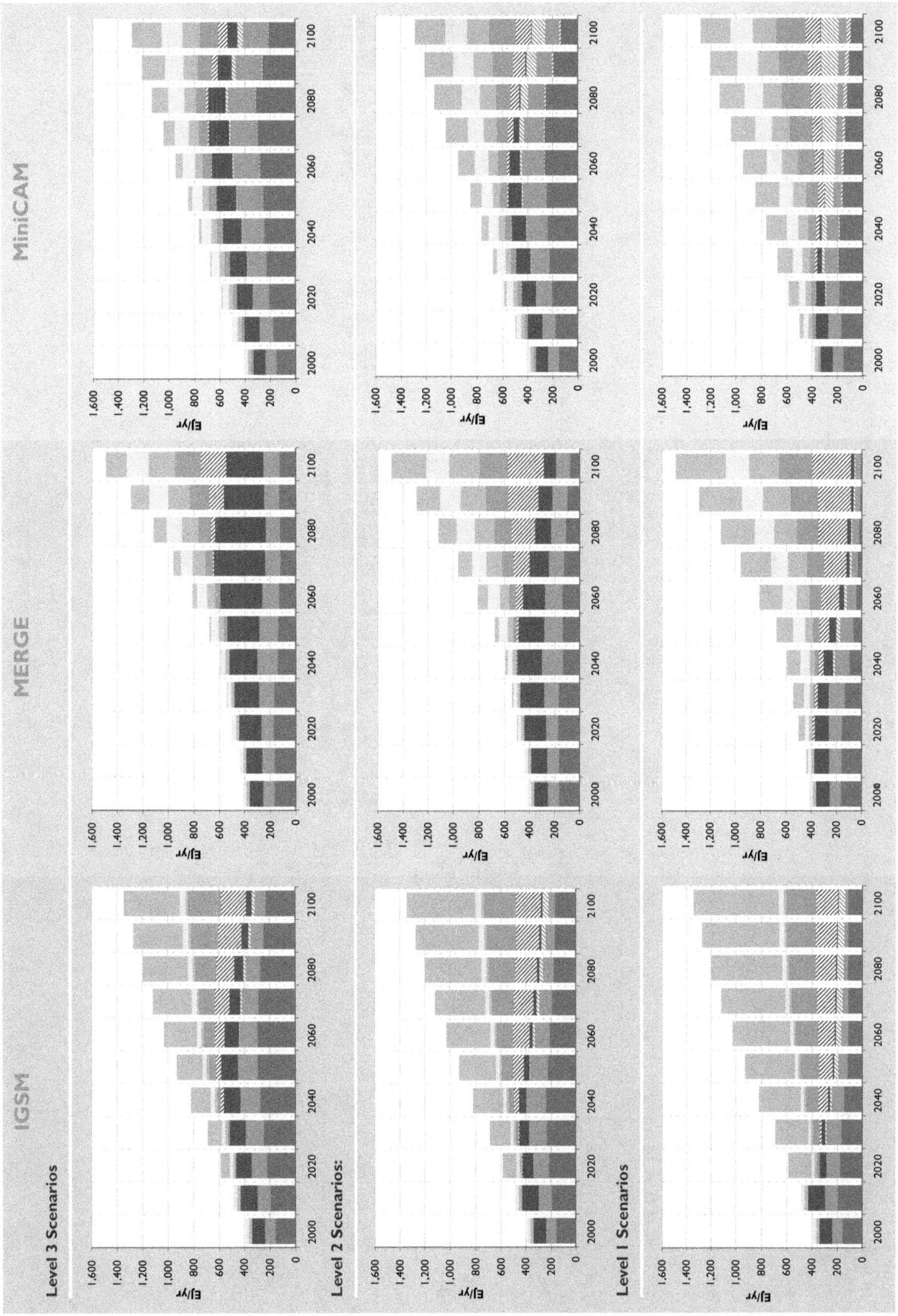
IGSM
MERGE
MiniCAM
Level 3 Scenarios
Level 2 Scenarios:
Level 1 Scenarios
EJ/yr
0
200
400
600
800
1,000
1,200
1,400
1,600
2000
2020
2040
2060
2080
2100

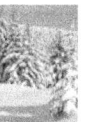

Figure 4.11. Change in Global Primary Energy Consumption by Fuel Across Stabilization Scenarios, Relative to Reference Scenarios (EJ/yr). The energy system is significantly transformed from the reference scenarios in the stabilization scenarios from all three modeling groups. The transformation begins later in the Level 3 and 4 stabilization scenarios, but would need to continue into the following century. The transformation includes reductions in energy consumption, increased use of carbon-free sources of energy (biomass, other renewables, and nuclear power), and the addition of CCS. The contribution of each of these varies among the models, reflecting different assumptions about economic viability, non-climate policy, and resource limits. [Notes. *i.* Oil consumption includes that derived from tar sands and oil shales, and coal consumption includes that used to produce synthetic liquid and gaseous fuels. *ii.* Primary energy consumption from nuclear power and non-biomass renewable electricity are accounted for at the average efficiency of fossil-fired electric facilities, which vary over time and across scenarios. This long-standing convention means that, all other things being equal, increasing efficiency of fossil-electric energy lowers the contribution to primary energy from these sources.]

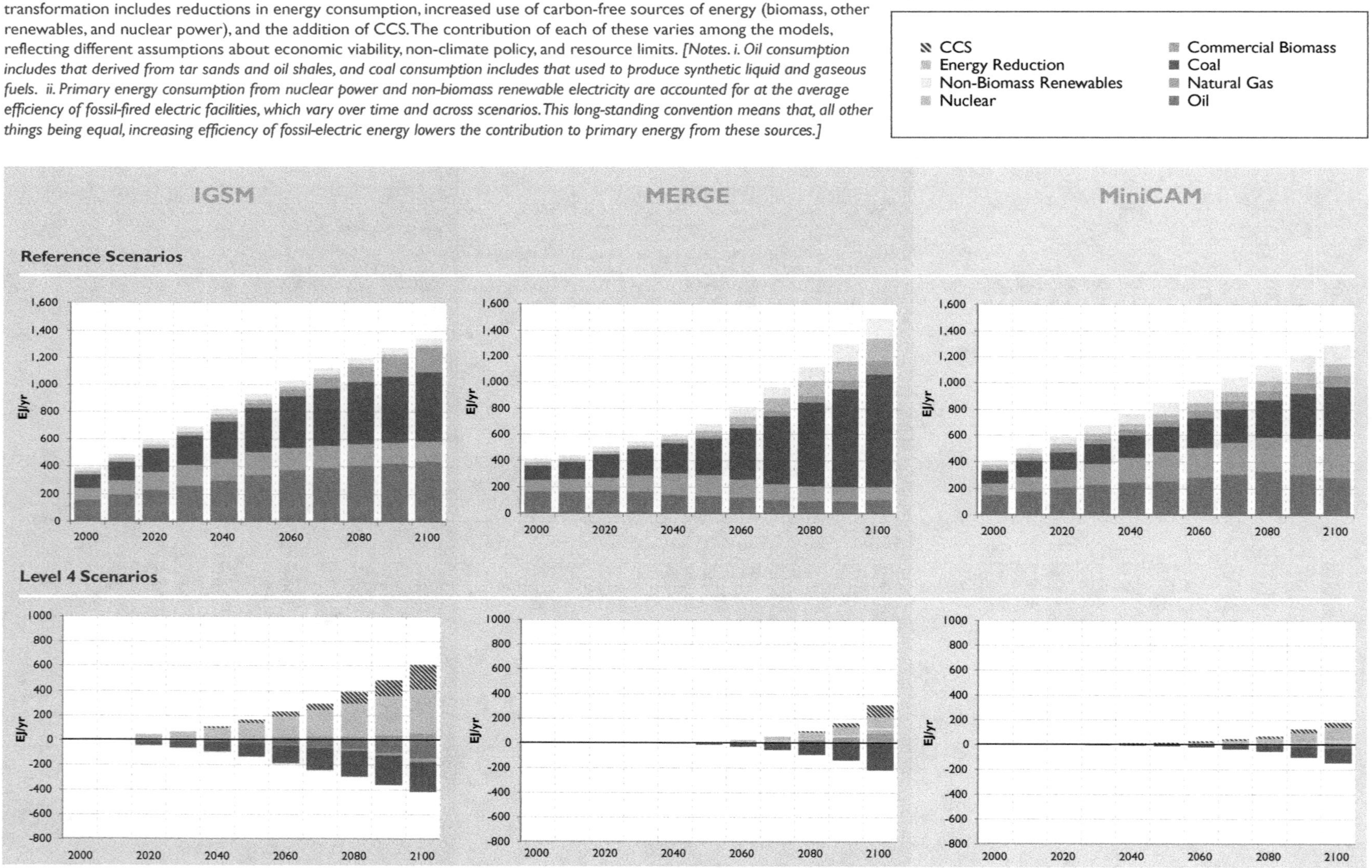

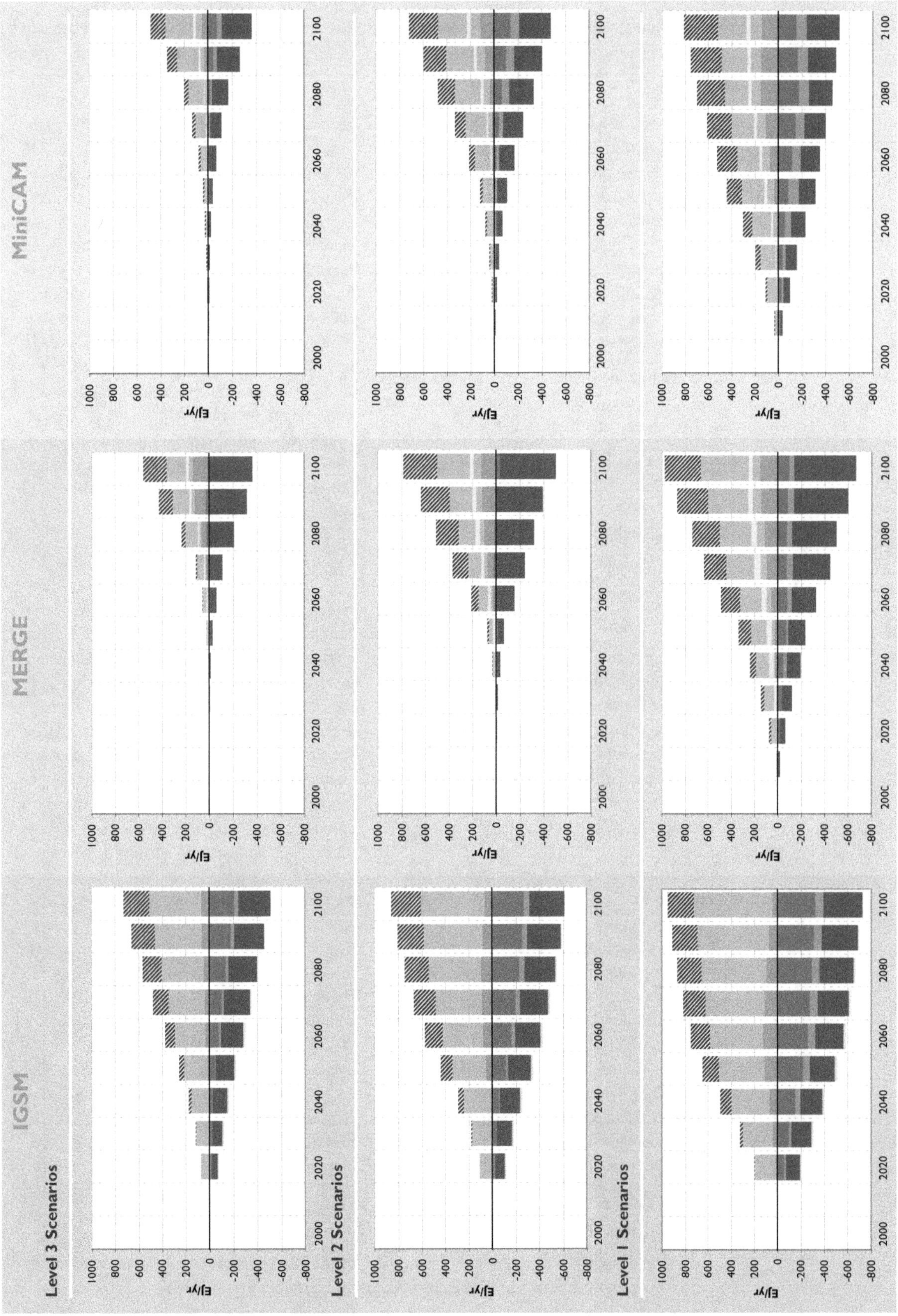

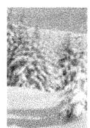

MiniCAM

MERGE

IGSM

Level 3 Scenarios

Level 2 Scenarios

Level 1 Scenarios

ture of these three models implies that technology deployments are a reflection of prices. Technologies are deployed up to the point where marginal cost is equal to price. For example, the prices of oil and carbon set the price at which bio-fuels compete. It is therefore possible to infer the marginal costs of bio-fuels when they first enter the market and how the marginal cost changes as the market expands.

It is worth reemphasizing that reductions in energy consumption are an important component of response at all stabilization levels. These reductions reflect a mix of three factors:

- Substitution of technologies that produce the same energy service with lower direct-plus-indirect carbon emissions

- Changes in the composition of final goods and services, shifting toward consumption of goods and services with lower direct-plus-indirect carbon emissions

- Reductions in the consumption of energy services.

This report does not attempt to quantify the relative contribution of each of these responses. Each of the models has a different set of technology options, different technology performance assumptions, and different model structures. Furthermore, no well defined protocol exists that can provide a unique attribution among these three general processes.

Changes in Global Electric Power Generation

Across the scenarios, stabilization leads to substantial changes in electricity-production technologies, although the MERGE and MiniCAM stabilization scenarios exhibit relatively little change in electricity production. Indeed, across the models, the relative reductions in electricity production under stabilization are lower than relative reductions in total primary energy consumption. One reason for this is that electricity price increases are smaller relative to those for direct fuel use because the fuel input, while important, is only part of the consumer cost of electricity. Also, the long-term cost of the transition to low and non-carbon-emitting sources is relatively smaller in electricity production than in the remaining sectors taken as an average.

There are substantial differences in the scale of global electricity production across the three reference scenarios, as shown in Chapter 3 and repeated at the top of Figure 4.12. Electricity production increases from about 50 EJ/yr in the year 2000 to between 230 EJ/yr (IGSM) to 310 EJ/yr (MiniCAM) by 2100. In all three reference scenarios, electricity becomes an increasingly important component of the global energy system, fueled by growing quantities of fossil fuels. Despite differences in the relative contribution of different fuel sources across the three reference scenarios, total production of electricity from fossil fuel rises from about 30 EJ/yr in 2000 to between 150 EJ/yr and 190 EJ/yr in 2100. Thus, the difference in total reference scenario electricity production among the models largely reflects differences in the deployment of non-fossil energy forms: bio-fuels, nuclear power, fuel cells, and other renewables such as wind, geothermal, and solar power.

The imposition of radiative forcing limits dramatically changes the electricity sector. Common characteristics of the stabilization scenarios across models are that CCS (with coal, gas, and, where present, oil-generated power) is deployed at a large scale by the end of the century and that use of coal without CCS declines and eventually is not viable. The IGSM scenarios, as has been noted, assume restrictions on the expansion of nuclear power, and other renewables are either resource limited (hydro power and electricity from bio-fuels) or become more costly to integrate into the grid as their share of electricity production rises because they are intermittent (wind and/or solar). Partly as a result, natural gas use in electricity production increases in the IGSM stabilization scenarios, especially in the nearer term before CCS becomes economically viable. In the MERGE stabilization scenarios, carbon-free technologies, including non-biomass renewables and nuclear, are viable and, thus, are favored over natural gas, the use of which falls relative to the reference scenario. In the MiniCAM stabilization scenarios, nuclear and non-biomass renewable energy technologies capture a larger share of the market. At the less stringent levels of stabilization, Level 3 and Level 4, additional bio-fuels are deployed in electricity production, and total electricity production declines. Under the most stringent stabilization level, commercial bio-fuels used in electricity production in the

Table 4.3. Global Annual CO_2 Capture and Storage in 2030, 2050, and 2100 for Four Stabilization Levels.

Stabilization Level	Year	Annual Global Carbon Capture and Storage (GtC/yr)		
		IGSM	MERGE	MiniCAM
Level 4	2030	0.01	0.00	0.09
	2050	0.44	0.00	0.15
	2100	4.12	2.31	0.72
Level 3	2030	0.05	0.00	0.10
	2050	0.83	0.00	0.19
	2100	4.52	4.79	2.75
Level 2	2030	0.12	0.00	0.13
	2050	1.96	0.44	0.38
	2100	4.97	6.63	5.56
Level 1	2030	0.37	0.66	0.82
	2050	2.76	2.24	2.95
	2100	4.44	7.17	6.23

Table 4.4. Global Cumulative CO_2 Capture and Storage in 2050 and 2100 for Four Stabilization Levels.

Stabilization Level	Year	Cumulative Global Carbon Capture and Storage (GtC)		
		IGSM	MERGE	MiniCAM
Level 4	2030	0.0	0.0	1.1
	2050	3.6	0.0	3.4
	2100	91.7	21.1	20.7
Level 3	2030	0.2	0.0	1.2
	2050	8.5	0.0	4.0
	2100	152.8	64.2	51.8
Level 2	2030	0.5	0.0	1.5
	2050	19.5	3.2	6.4
	2100	208.0	187.7	144.2
Level 1	2030	1.8	7.4	6.9
	2050	36.7	32.4	43.0
	2100	230.6	272.5	278.0

MiniCAM stabilization scenarios are diverted to the transportation sector, and use in electricity production actually declines relative to the reference toward the end of the century. In all of the IGSM scenarios, bio-fuels are used preferentially for transportation rather than for electricity generation. The difference between MiniCAM and IGSM scenarios in this regard is in part a reflection of the higher fuel prices in the IGSM scenarios discussed in Section 4.6.3.

All modeling groups assumed that CO_2 could be captured and stored in secure repositories, and as noted, in all scenarios CCS becomes a large-scale activity. Annual capture quantities are shown in Table 4.3. CCS is always one of the largest single changes in the electricity production system in response to stabilization, as can be seen in Figure 4.13. As with mitigation in general, CCS starts relatively modestly in all the scenarios, but grows to large levels. The total

Figure 4.12. Global Electricity Production by Fuel Across Scenarios (EJ/yr). Global electricity production would need to be transformed to meet the four stabilization levels. CCS is important in the scenarios from all three modeling groups; thus, while coal use is reduced in all the stabilization scenarios relative to the reference scenarios, it remains an important fuel for electricity production. Use of CCS is the main supply response in the IGSM stabilization scenarios, in part because nuclear power is limited by assumption to reflect non-climate policy concerns. Nuclear power and renewable electricity sources play a larger role in the MERGE and MiniCAM stabilization scenarios.

Legend:
- Non-Biomass Renewables
- Nuclear
- Commercial Biomass
- Coal: w/ CCS
- Coal: w/o CCS
- Natural Gas: w/ CCS
- Natural Gas: w/o CCS
- Oil: w/ CCS
- Oil: w/o CCS

IGSM, MERGE, MiniCAM — Reference Scenarios and Level 4 Scenarios

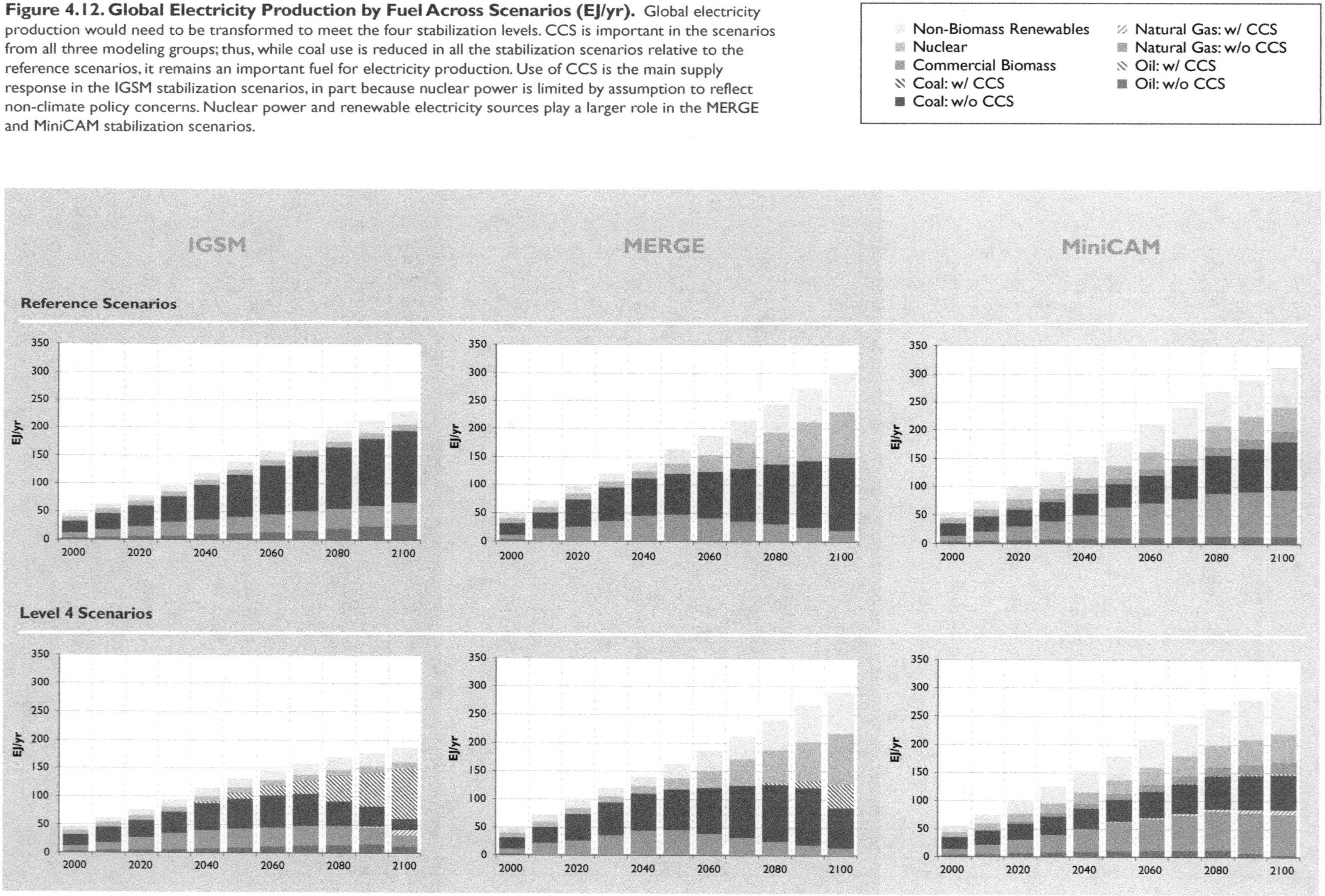

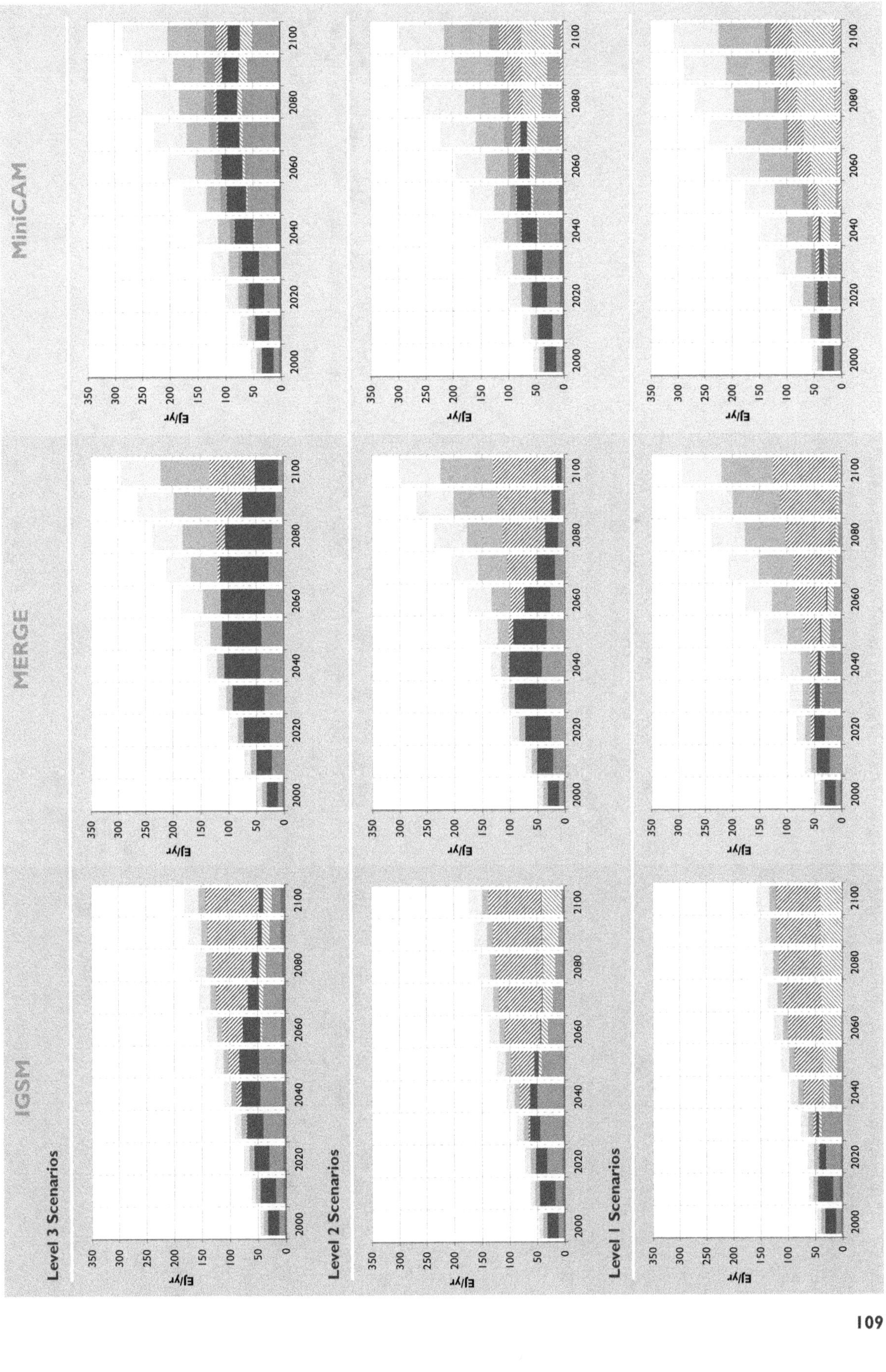

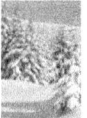

IGSM

MERGE

MiniCAM

Level 3 Scenarios

Level 2 Scenarios

Level 1 Scenarios

Figure 4.13. Changes in Global Electricity Production by Fuel Across Stabilization Scenarios, Relative to Reference Scenarios (EJ/yr). There are multiple electricity technology options that could be competitive in the future, and different assumptions about their relative economic viability, reliability, and resource availability lead to different scenarios for the global electricity sector in reference and stabilization scenarios across the models. In the IGSM reference scenario, there is relatively little change in the fuel mix in the electricity sector, with continued reliance on coal. In the MERGE and MiniCAM reference scenarios, there are large transformations from the present. In the stabilization scenarios from all three modeling groups, large changes relative to the reference scenario are required to meet the stabilization levels. Under less stringent stabilization levels, many of these changes would be pushed into the next century. In most cases, the relative proportion of electricity in energy consumption increases in the stabilization scenarios, so the relative reductions in electricity production are generally smaller than for primary energy.

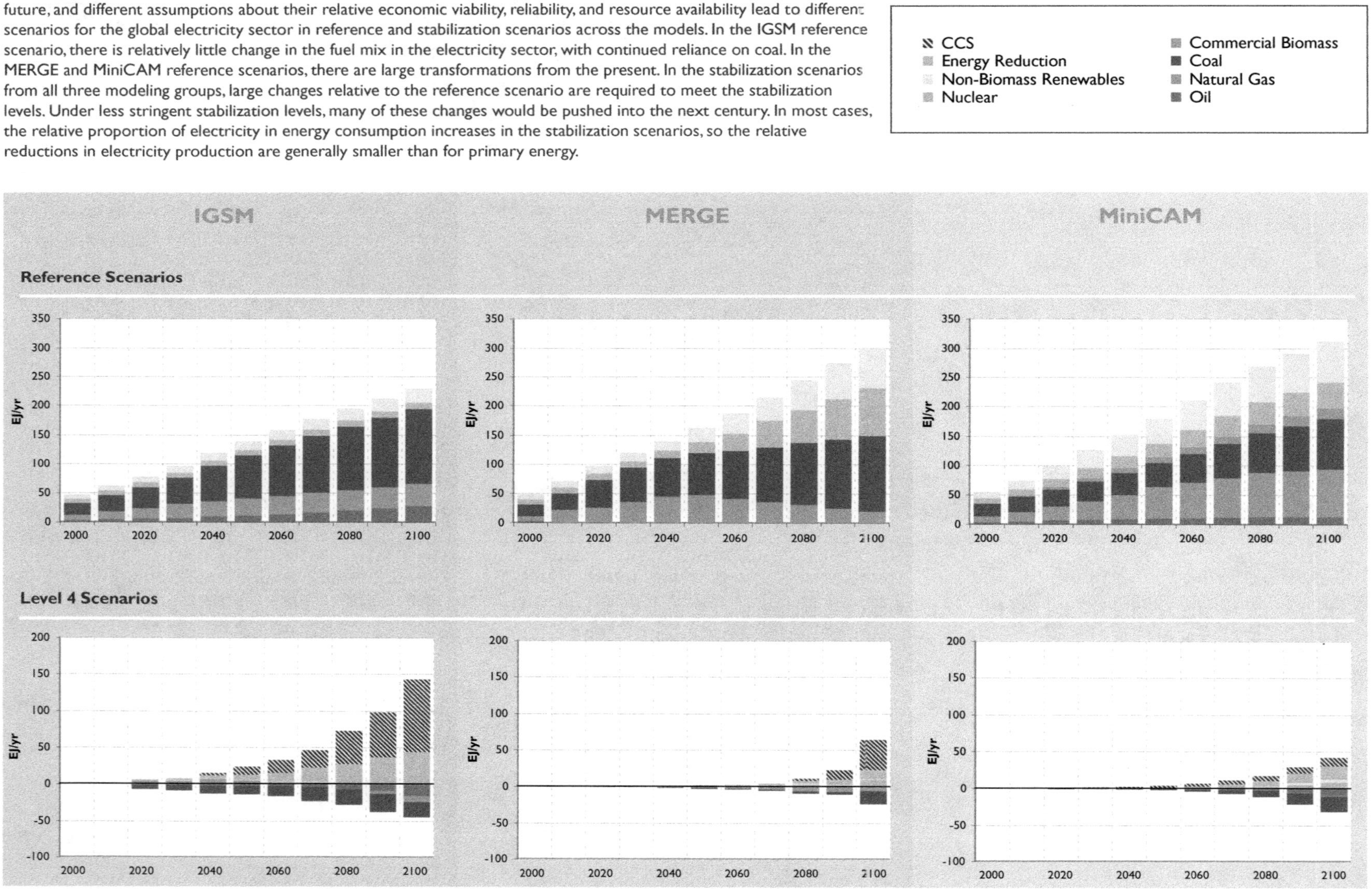

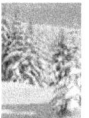

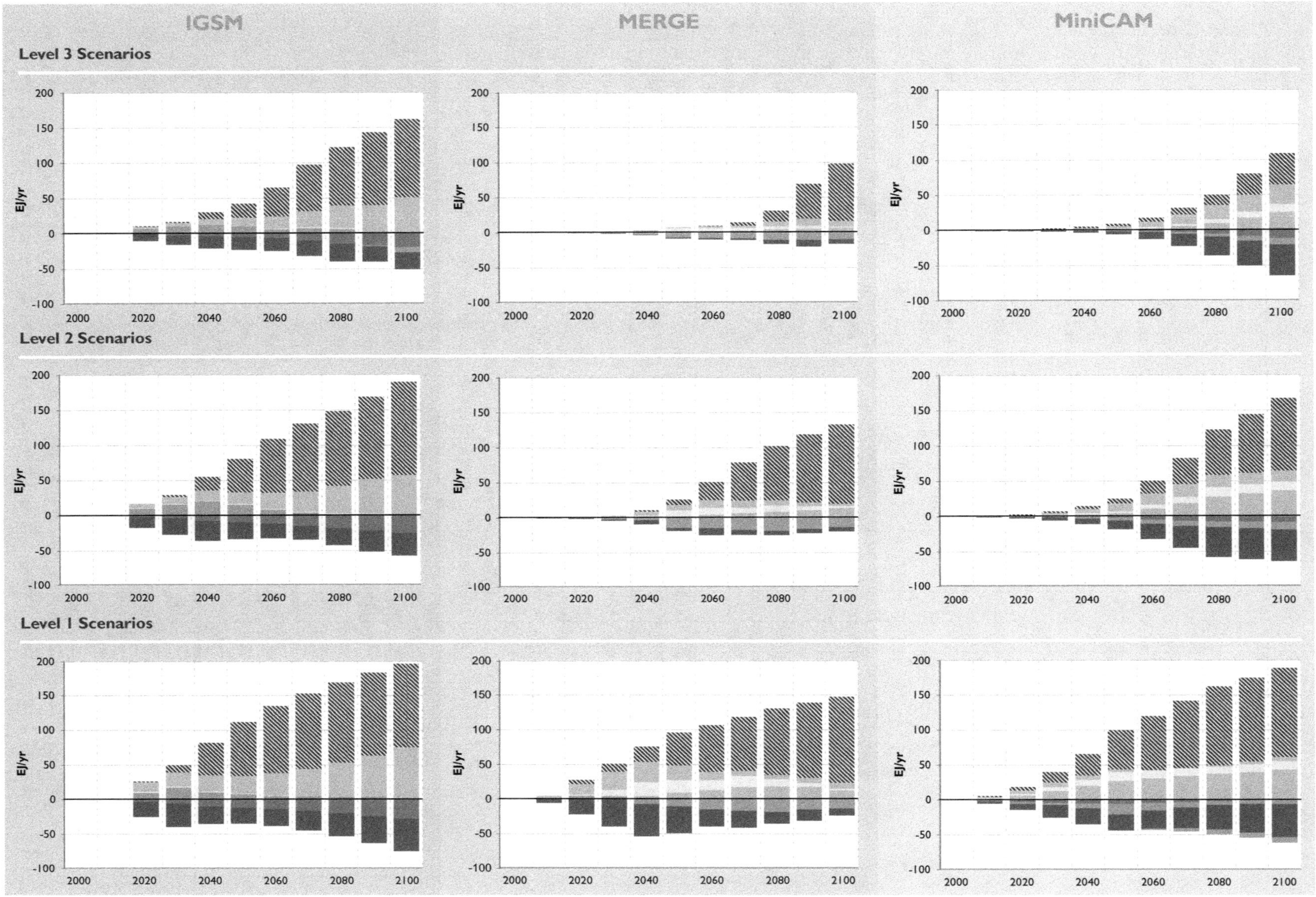
IGSM
MERGE
MiniCAM
Level 3 Scenarios
Level 2 Scenarios
Level 1 Scenarios
EJ/yr
200
150
100
50
0
-50
-100
2000
2020
2040
2060
2080
2100

storage over the century is recorded in Table 4.4, spanning a range from 20 GtC to 90 GtC for Level 4 and 230 GtC to 280 GtC for Level 1. The modeling groups did not report either location of storage sites for CO_2 or the nature of the storage reservoirs, but these scenarios are within the range of the estimates of global geologic reservoir capacity (Edmonds et al. 2001, Dooley et al. 2004).

Deployment rates for CCS depend on a variety of circumstances, including capture cost, new plant construction versus retrofitting for existing plants, the scale of power generation, the price of fuel inputs, the cost of competing technologies, and the level of the CO_2 price. It is clear that the constraints on radiative forcing considered in these scenarios are sufficiently stringent that, if CCS is available at a cost and performance similar to that considered in these scenarios, and that it successfully navigates other potential obstacles to widespread deployment, it could be a crucial component of future power generation.

Yet CCS is hardly ordinary today. Geologic storage is largely confined to experimental sites or enhanced oil and gas recovery. There are as yet no clearly defined institutions or accounting systems to reward such technology in emissions control agreements, and long-term liability for stored CO_2 has not been determined. All of these issues and more must be resolved before CCS could deploy on the scale envisioned in these stabilization scenarios. If CCS were unavailable, the effect would be to increase the cost of achieving stabilization in all of the scenarios. These scenarios tend to favor CCS, but that tendency could easily change with different assumptions about technologies such as nuclear power that are well within the range of uncertainty about future costs and the policy environment. Nuclear power carries with it issues of safety, waste, and proliferation. Thus, the viability of both CCS and nuclear power depends on regulatory and public acceptance issues. For example, global nuclear power in the reference scenarios ranges from about 1½ times current levels (if non-climate concerns such as safety, waste, and proliferation constrain its growth as is the case in one reference scenario), to an expansion of almost an order of magnitude assuming relative economics as the only constraint.

Absent CCS and nuclear power, these models would need to deploy other emissions reduction options that could potentially be more costly, or would need to assume large breakthroughs in cost, performance, and reliability. This study has not attempted to quantify the increase in costs or the reorganization of the energy system that would be required to achieve stabilization without CCS. This sensitivity is an important item in the agenda of future research.

Changes in Energy Patterns in the United States

Changes in U.S. energy patterns are similar to those observed for the world in general. This reflects the facts that the mitigation policy is implemented globally, there are international markets in fuels, each model makes most technologies globally available over time, and the U.S. primary energy consumption in 2000 represented roughly a quarter of the world total.

Changes in the U.S. energy system are modest for stabilization Level 4, but even with this loose constraint, significant changes begin upon implementation of the stabilization policy (the first period shown is 2020) in the IGSM Level 4 scenario (Figure 4.14 and Figure 4.15). Near-term changes are more modest in the MERGE and MiniCAM Level 4 scenarios. At more stringent stabilization levels, the changes are more substantial in the scenarios from all three modeling groups. In the Level 1 scenarios, the reduction is in U.S. primary energy consumption ranges from 8 EJ/yr to over 25 EJ/yr in 2020.

Near-term changes in the U.S. energy system vary more among models than the long-term adjustments. While oil consumption declines at higher carbon prices for all the models and all stabilization levels, near-term changes in oil consumption do not follow a consistent pattern. However, there is no ambiguity regarding the effect on coal consumption, which declines relative to the reference scenario in all stabilization scenarios for all models in all time periods. Similarly, total primary energy consumption declines in all the stabilization scenarios. Nuclear power, commercial biomass, and other renewable energy forms are advantaged with at least one of them always deployed to a greater extent in stabilization scenarios than in the reference scenario. The particular form and timing of expanded development varies across models.

The stabilization scenarios from the three modeling groups exhibit different energy sector responses reflecting differences in underlying reference scenarios and technology assumptions. The largest change in the U.S. energy system in the IGSM stabilization scenarios is the reduction in total primary energy consumption augmented by an expansion in the use of commercial biomass fuels and deployment of CCS. Similarly, the largest change in the MERGE scenarios is the reduction in total primary energy consumption augmented by deployment of CCS and bioenergy. The MiniCAM stabilization scenarios also exhibit reductions in primary energy consumption and increases in nuclear power, along with smaller additions of commercial biomass and other renewable energy forms. The adjustment of the U.S. electric sector to the various stabilization levels is similar to that for the world electricity sector. (Figure 4.16 and Figure 4.17).

IMPLICATIONS FOR AGRICULTURE, LAND-USE, AND TERRESTRIAL CARBON

In the stabilization scenarios, increased use is made of biomass energy crops, the contribution of which is ultimately limited by competition with agriculture and forestry. Two of the modeling groups employed explicit agriculture-land-use models to represent this competition and represent land constraints on the use of bio-energy. In the scenarios from one modeling group, increased use of bio-energy at more stringent stabilization levels leads to substantial land use change emissions as previously unmanaged lands are shifted to biomass production.

The three modeling groups employed different approaches to the treatment of the terrestrial carbon cycle, ranging from a simple neutral biosphere model to a state-of-the-art terrestrial carbon-cycle model. In two of the models, a CO_2 fertilization effect plays a significant role. As stabilization levels become more stringent, CO_2 concentrations decline and terrestrial carbon uptake declines, with implications for emissions mitigation in the energy sector. Despite the differences across the modeling groups' treatments of the terrestrial carbon cycle, the aggregate behavior of the carbon cycles across models is similar.

In the stabilization scenarios, the cost of using fossil fuels and emitting CO_2 rises, providing an increasing motivation for the production and transformation of bioenergy, as shown in Figure 4.18. In all of the stabilization scenarios, production begins earlier and produces a larger share of global energy as the stabilization level becomes more stringent. Under less stringent stabilization levels, production of bio-crops is lower in the second half of the century in the MERGE and MiniCAM scenarios than in the IGSM scenarios. Differences between the models with respect to biomass deployment are not simply due to different treatments of agriculture and land use but also result from the full suite of competing technologies and behavior assumptions.

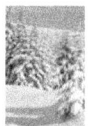

Although total land areas allocated to bioenergy crops are not reported in these scenarios, the extent of land areas engaged in the production of energy becomes substantial. This is possible only if appropriate land is available, which hinges on future productivity increases for other crops and the potential of bioenergy crops to be grown on lands that are less suited for food, pasture, and forests. In both the MiniCAM and IGSM scenarios – MiniCAM and IGSM are the two models with agriculture and land-use submodels – demands on land for bio-fuels cause land prices to increase substantially as compared with the reference scenarios because of competition with other agricultural demands.

Stabilization scenarios limit the rise in CO_2 concentrations and reduce the CO_2 fertilization effect below that in the reference scenarios, which in turn leads to smaller CO_2 uptake by the ter-

Figure 4.14. U.S. Primary Energy Consumption by Fuel Across Scenarios (EJ/yr). U.S. primary energy consumption under the four stabilization levels differs considerably among the three models. All the scenarios exhibit a diverse energy mix throughout the century, although the IGSM scenarios include relatively less nuclear power and non-biomass renewables than the other models. The relative contributions of different technologies over the course of the century depend on the specific cost and performance characteristics of the competing technologies represented in the scenarios. *[Notes. i. Oil consumption includes that derived from tar sands and oil shales, and coal consumption includes that used to produce synthetic liquid and gaseous fuels. ii. Primary energy consumption from nuclear power and non-biomass renewable electricity are accounted for at the average efficiency of fossil-fired electric facilities, which vary over time and across scenarios. This long-standing convention means that, all other things being equal, increasing efficiency of fossil-electric energy lowers the contribution to primary energy from these sources.]*

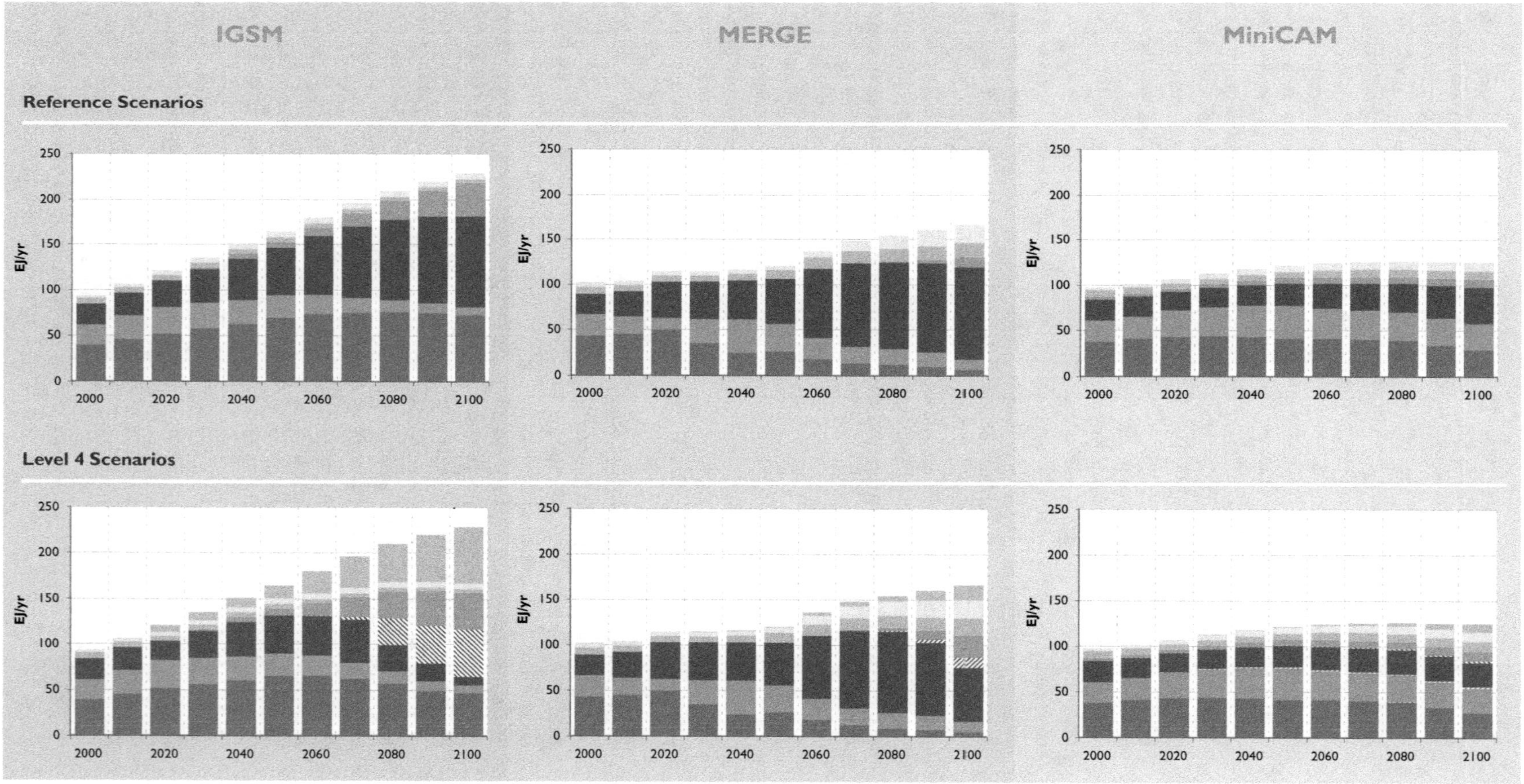

Legend:
- Non-Biomass Renewables
- Nuclear
- Commercial Biomass
- Coal: w/ CCS
- Coal: w/o CCS
- Natural Gas: w/ CCS
- Natural Gas: w/o CCS
- Oil: w/ CCS
- Oil: w/o CCS
- Energy Reduction

IGSM — Reference Scenarios
MERGE
MiniCAM

Level 4 Scenarios

EJ/yr

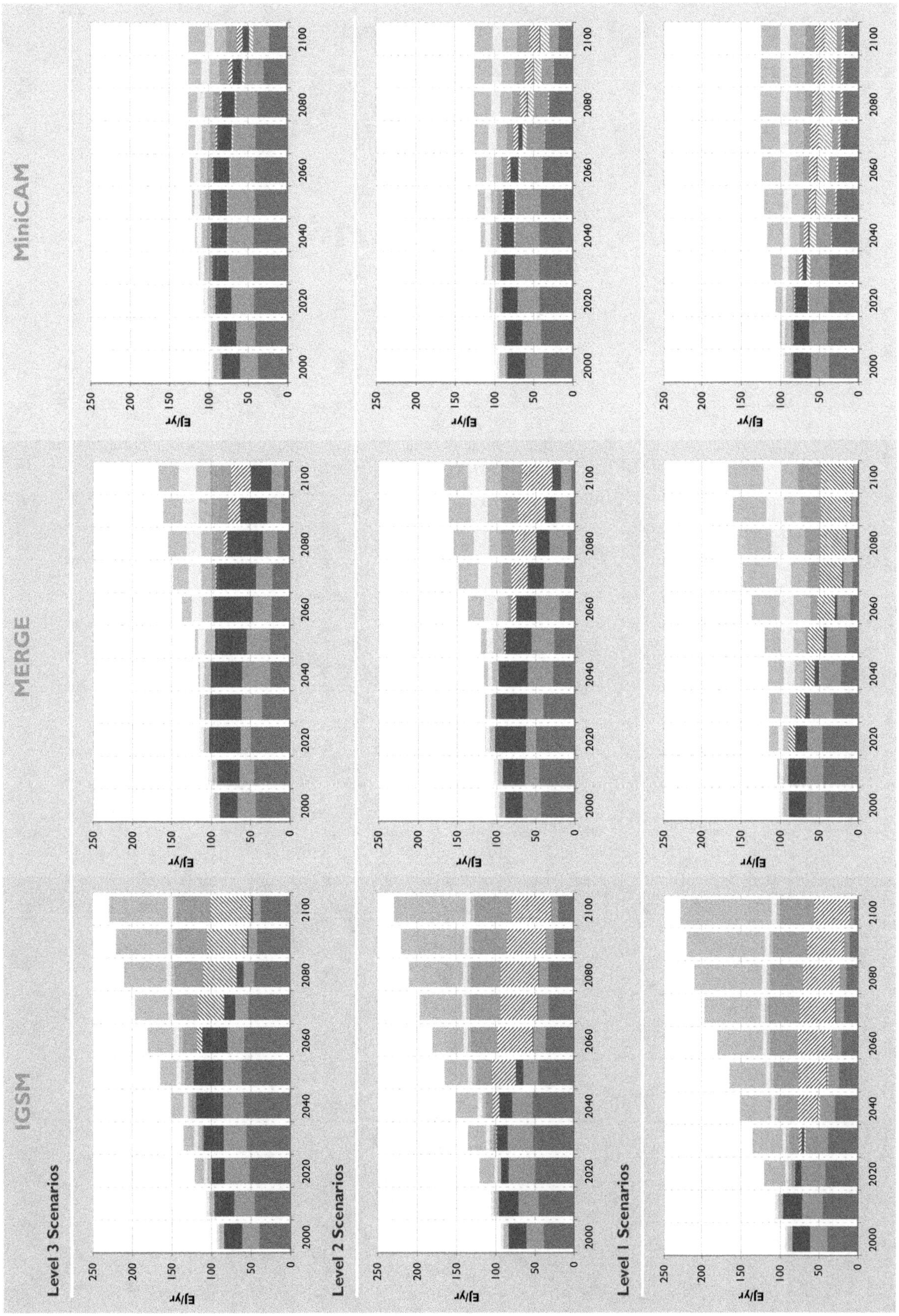

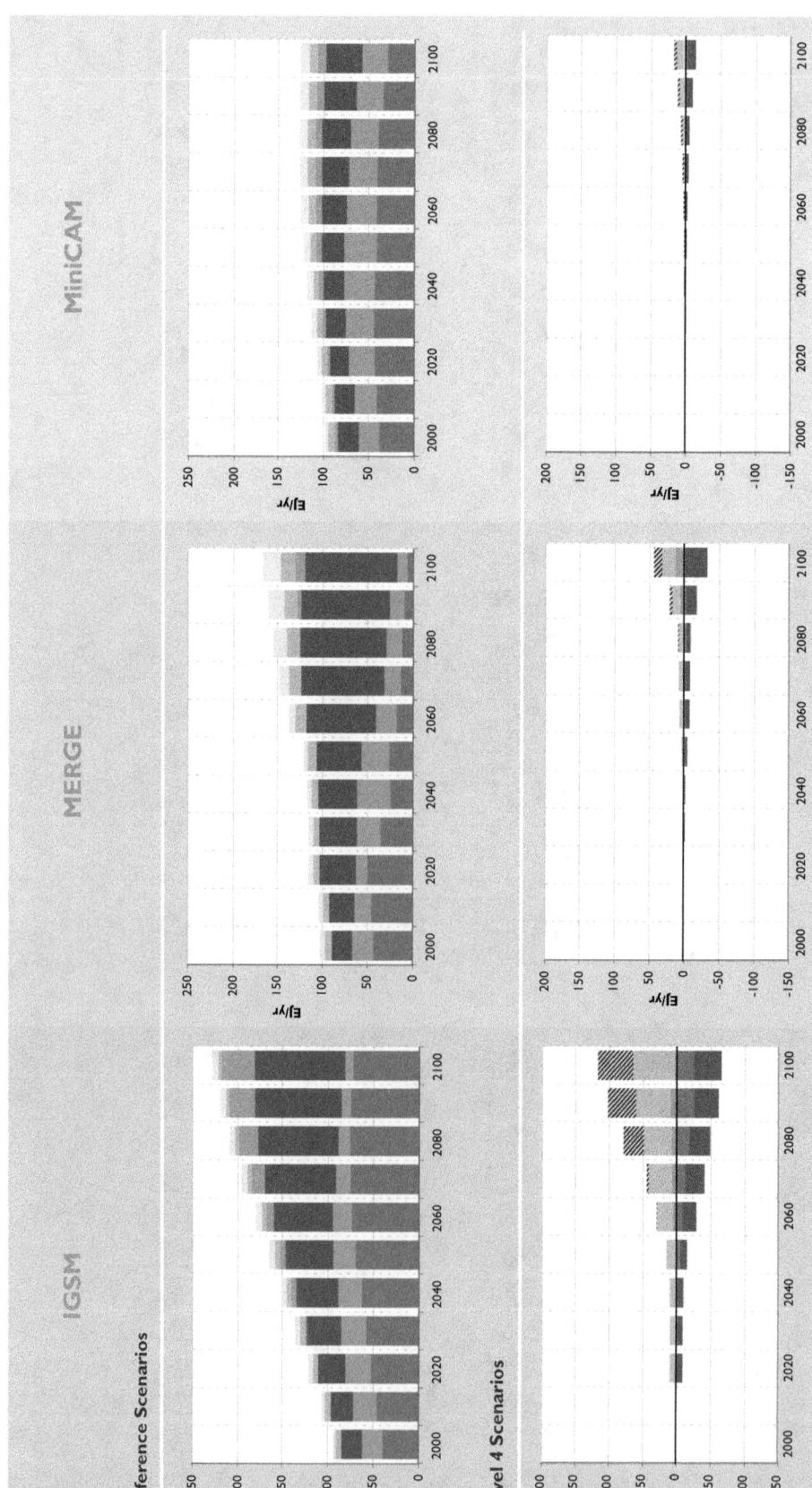

Figure 4.15. Changes in U.S. Primary Energy Consumption by Fuel Across Stabilization Scenarios, Relative to Reference Scenarios (EJ/yr). The transformations in the U.S. energy system in the stabilization scenarios are similar to those for the global energy system. Although it is not illustrated in this figure, one difference is the transformation from conventional oil and gas to synthetic fuel production derived from shale oil or coal. The IGSM reference scenario includes heavy use of shale oil with some coal gasification, whereas the MERGE reference scenario is based more heavily on synthetic liquid and gaseous fuels derived from coal. The MiniCAM reference scenario includes moderate levels of both. [Notes. i. Oil consumption includes that derived from tar sands and oil shales, and coal consumption includes that used to produce synthetic liquid and gaseous fuels. ii. Primary energy consumption from nuclear power and non-biomass renewable electricity are accounted for at the average efficiency of fossil-fired electric facilities, which vary over time and across scenarios. This long-standing convention means that, all other things being equal, increasing efficiency of fossil-electric energy lowers the contribution to primary energy from these sources.]

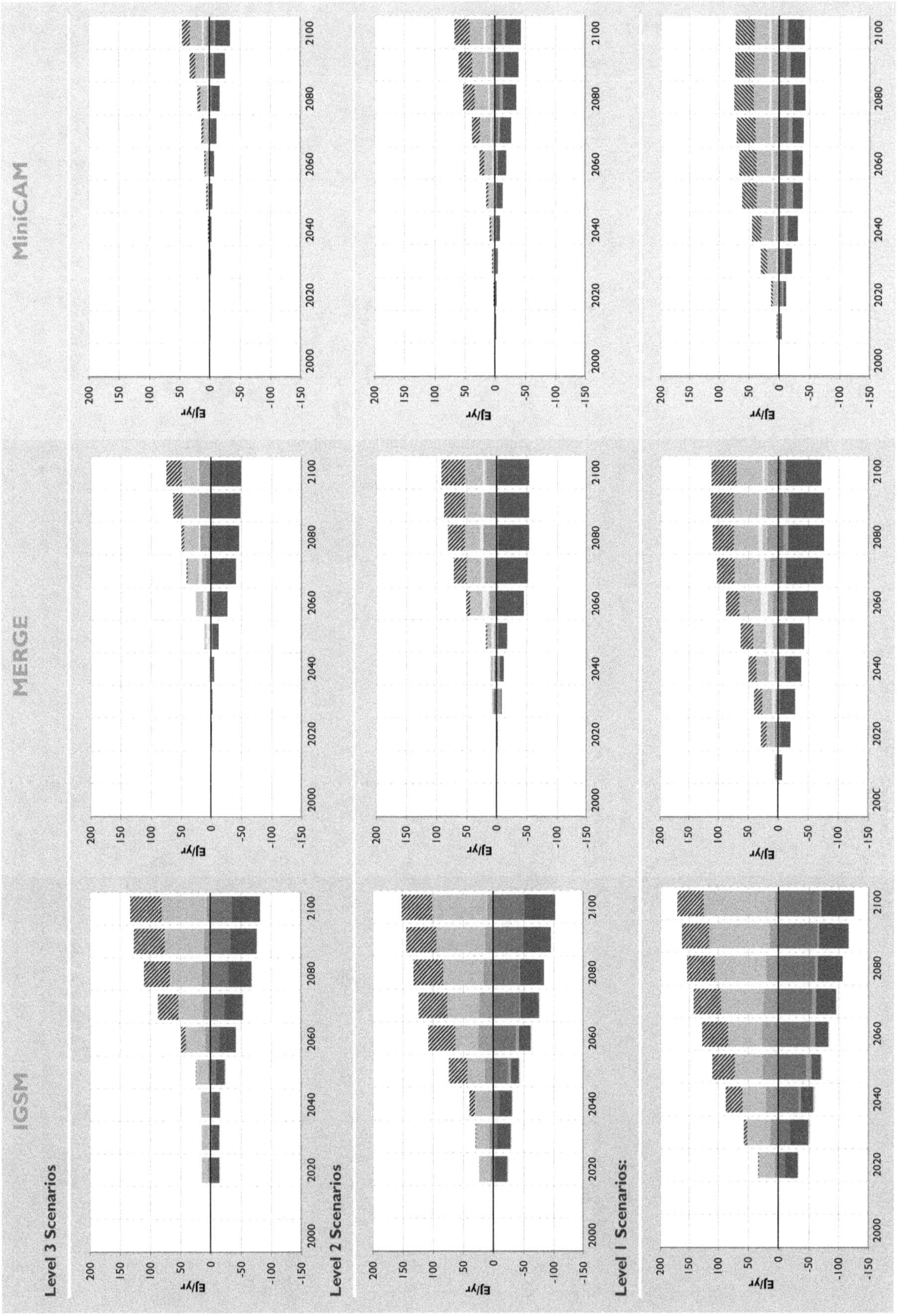

MiniCAM

EJ/yr

EJ/yr

EJ/yr

MERGE

EJ/yr

EJ/yr

EJ/yr

IGSM

Level 3 Scenarios

EJ/yr

Level 2 Scenarios

EJ/yr

Level 1 Scenarios:

EJ/yr

Figure 4.16. U.S. Electricity Production by Fuel Across Scenarios (EJ/yr). In these scenarios, U.S. electricity production sources and technologies are substantially transformed to meet the four stabilization levels. CCS figures in all the stabilization scenarios, but the contribution of other sources and technologies as well as the total production electricity differs substantially among the three models.

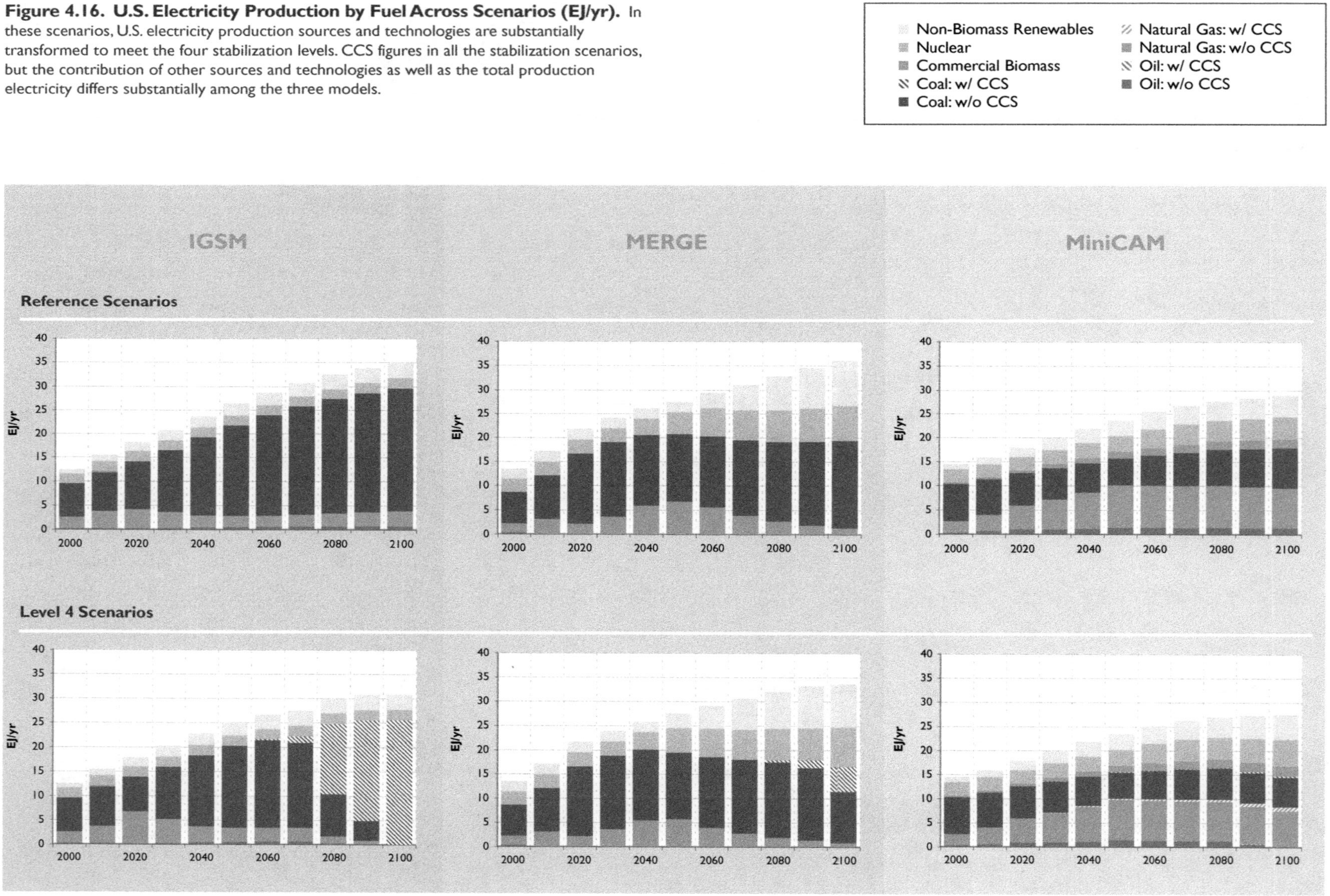

Legend:
- Non-Biomass Renewables
- Nuclear
- Commercial Biomass
- Coal: w/ CCS
- Coal: w/o CCS
- Natural Gas: w/ CCS
- Natural Gas: w/o CCS
- Oil: w/ CCS
- Oil: w/o CCS

IGSM

MERGE

MiniCAM

Reference Scenarios

Level 4 Scenarios

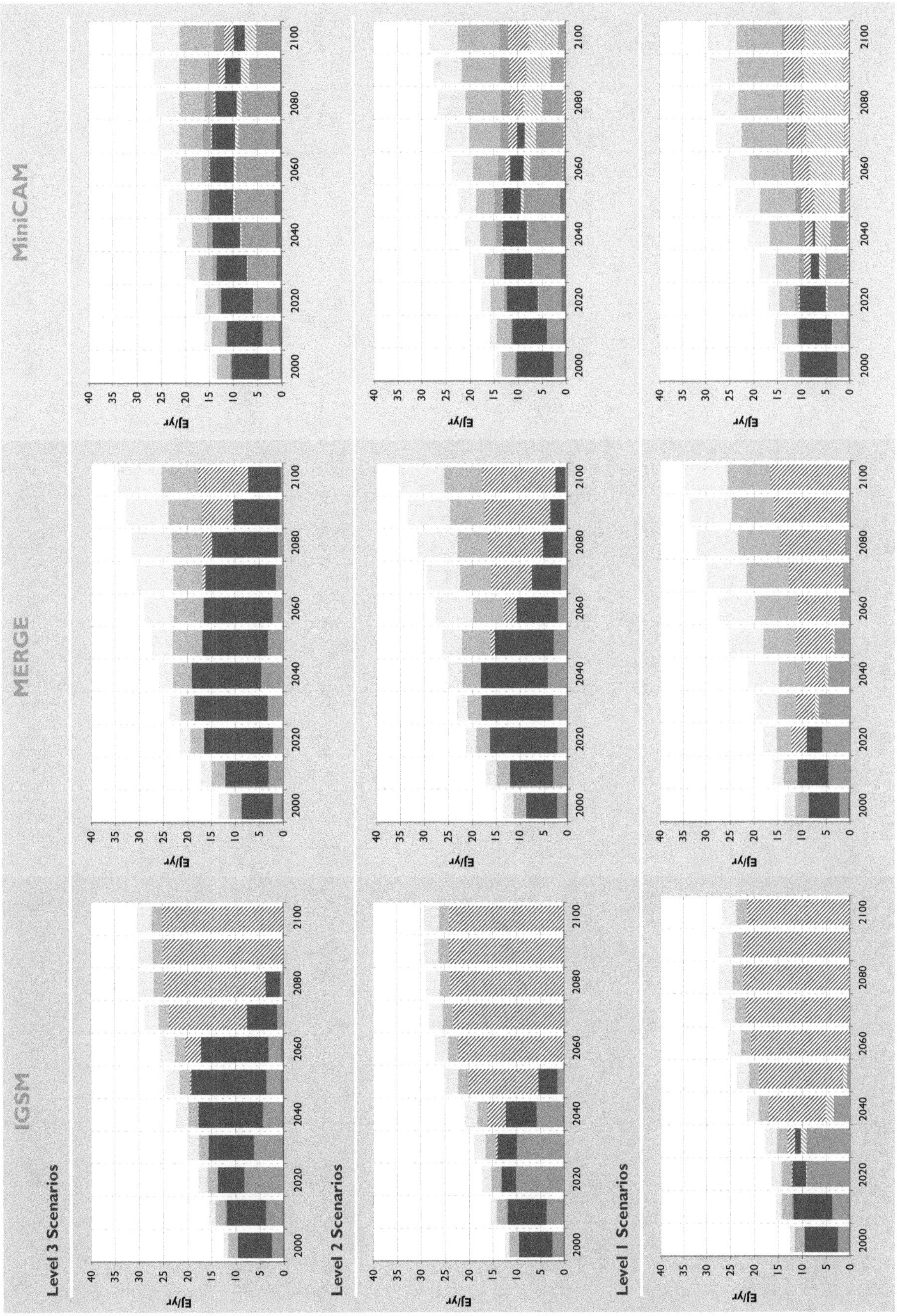

MiniCAM

EJ/yr

40 35 30 25 20 15 10 5 0

2000 2020 2040 2060 2080 2100

EJ/yr

40 35 30 25 20 15 10 5 0

2000 2020 2040 2060 2080 2100

EJ/yr

40 35 30 25 20 15 10 5 0

2000 2020 2040 2060 2080 2100

MERGE

EJ/yr

40 35 30 25 20 15 10 5 0

2000 2020 2040 2060 2080 2100

EJ/yr

40 35 30 25 20 15 10 5 0

2000 2020 2040 2060 2080 2100

EJ/yr

40 35 30 25 20 15 10 5 0

2000 2020 2040 2060 2080 2100

IGSM

Level 3 Scenarios

EJ/yr

40 35 30 25 20 15 10 5 0

2000 2020 2040 2060 2080 2100

Level 2 Scenarios

EJ/yr

40 35 30 25 20 15 10 5 0

2000 2020 2040 2060 2080 2100

Level 1 Scenarios

EJ/yr

40 35 30 25 20 15 10 5 0

2000 2020 2040 2060 2080 2100

Figure 4.17. Change in U.S. Electricity Production by Fuel Across Stabilization Scenarios, Relative to Reference Scenarios (EJ/yr).

Transformation of the U.S. electricity sector in these scenarios implies increasing use of low- or zero-carbon technologies, such as renewable electricity sources, nuclear power, and fossil generation with CCS, and decreasing use of fossil fuel technologies that freely emit CO_2 to the atmosphere. Natural gas use increases in the early part of the century in several stabilization scenarios as a lower carbon substitute for coal-fired electricity. In most cases, the relative proportion of electricity in energy consumption increases in the stabilization scenarios, so the relative reductions in electricity production are generally smaller than for primary energy. In one

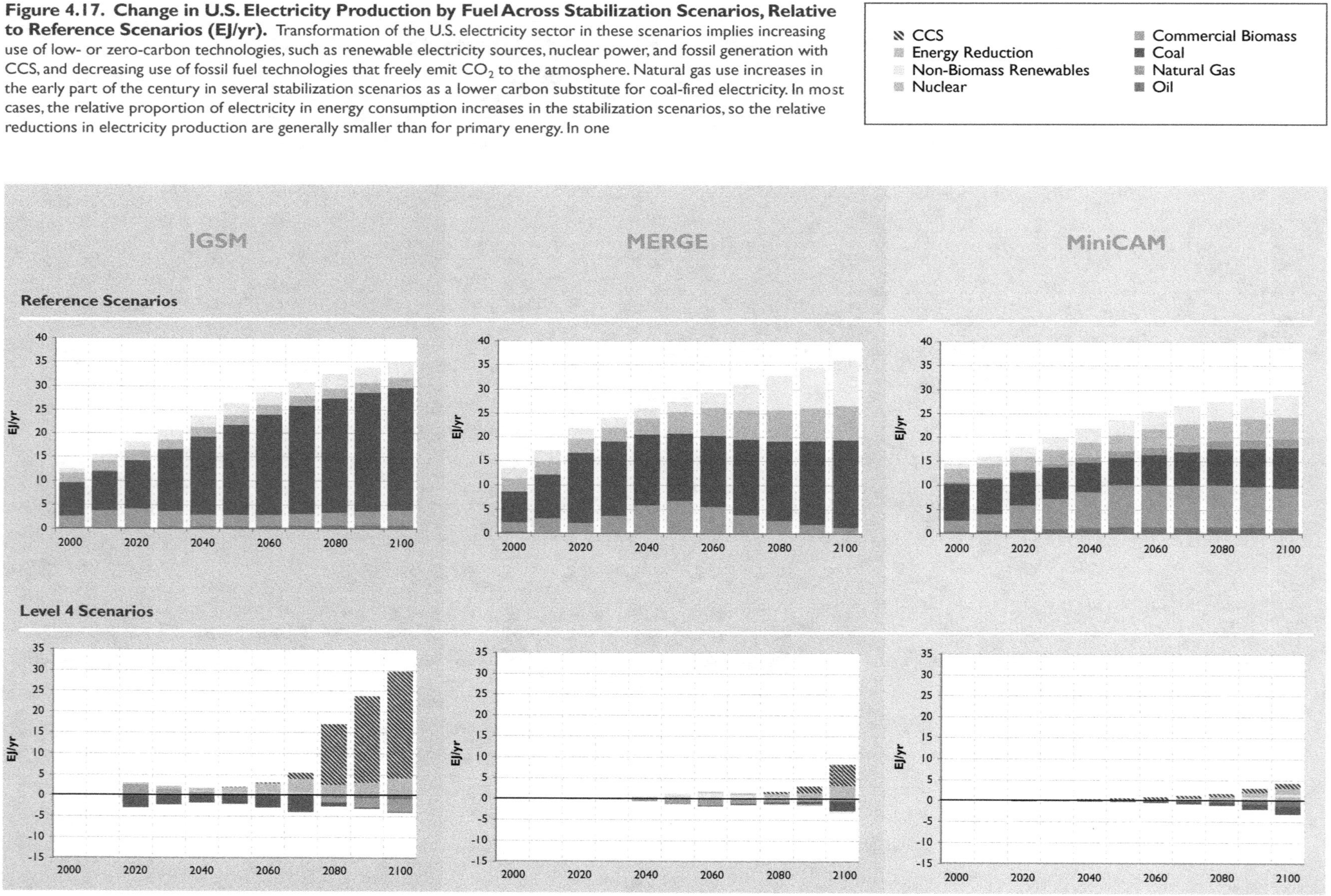

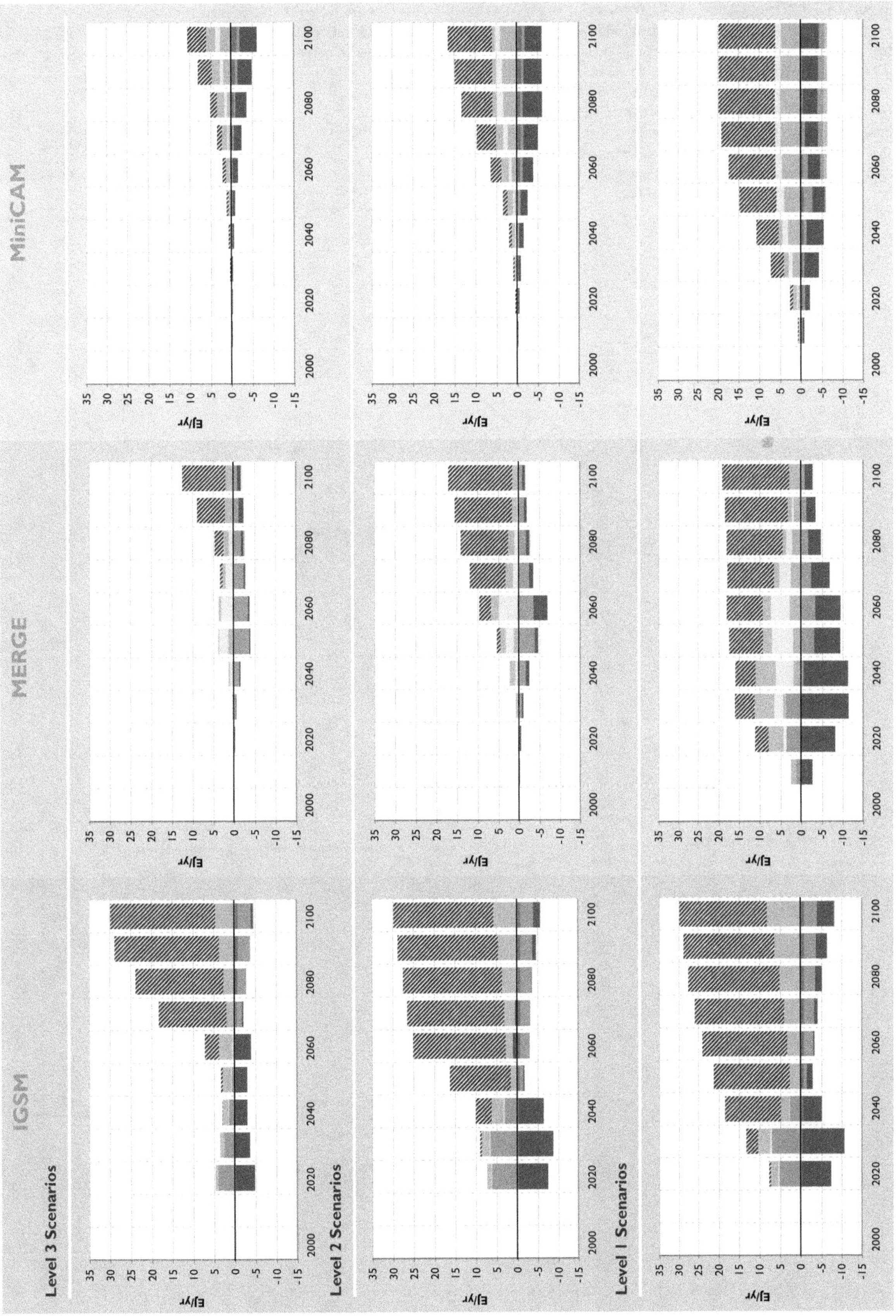

Figure 4.18. Global and U.S. Commercial Biomass Production Across Scenarios. Scenarios of the potential for commercial biomass production for the world and the U.S. are similar in magnitude and behavior among the models. Commercial biomass production increases over time in the reference scenarios due in large part to technological improvements in bioenergy crop production and increasing demand for liquid fuels. Stabilization increases the demand for bioenergy crops, causing production to increase more rapidly and to reach higher levels than in the reference scenarios. Dramatic growth in bioenergy crop production raises important issues about the attendant increases in the land that is devoted to these crops, including competition with other agricultural crops, encroachment into unmanaged lands, and water and other resource and environmental impacts.

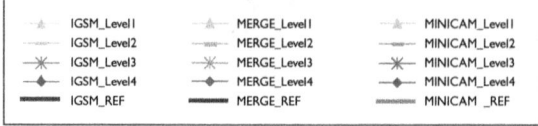

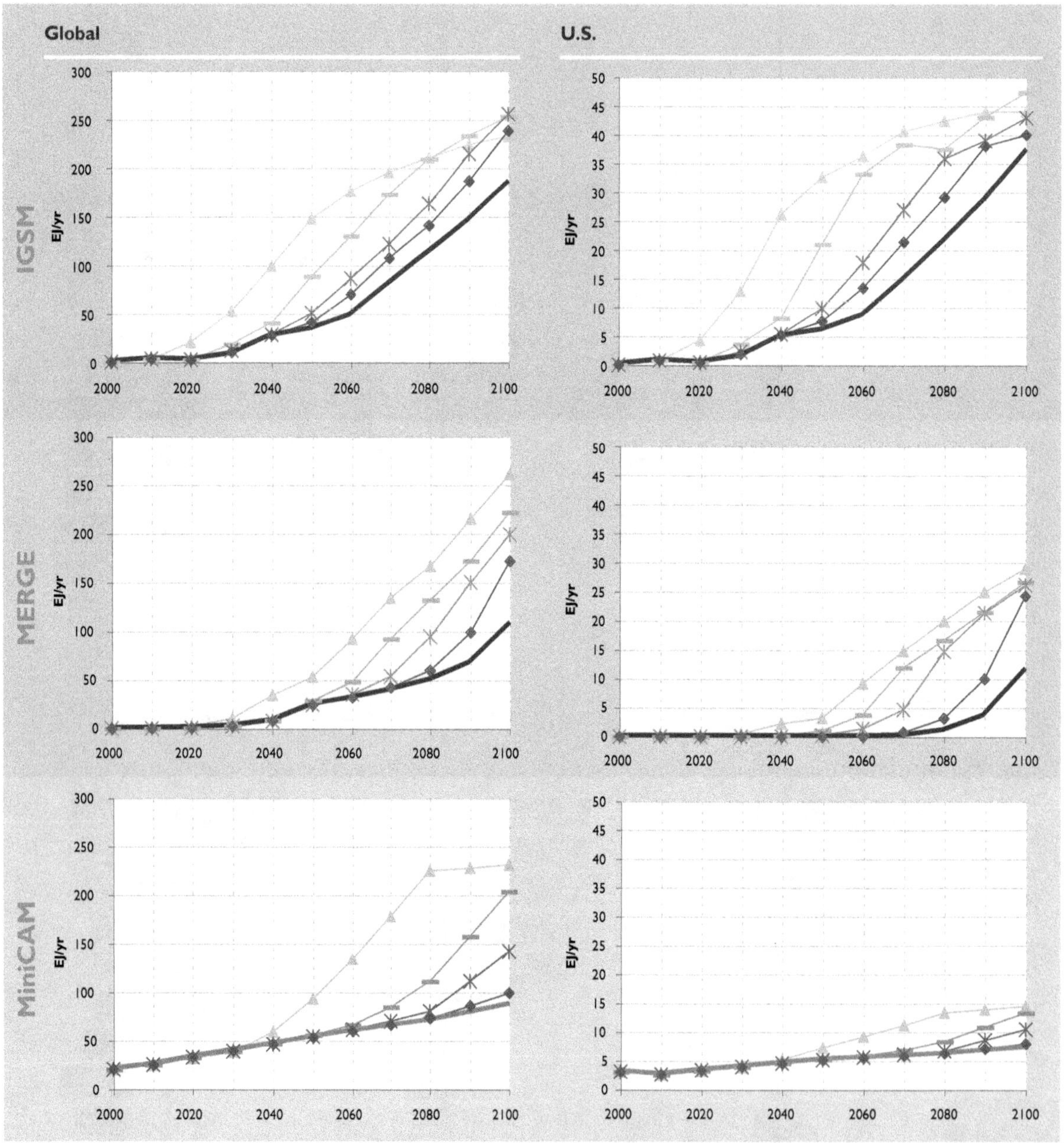

Figure 4.19. Net Terrestrial Carbon Emissions Across Scenarios (GtC/yr). Net terrestrial carbon emissions to the atmosphere, under reference and stabilization levels, reflect differences in the model structures for processes that remain highly uncertain. The MERGE scenarios are based on the assumption of a neutral biosphere. The IGSM and MiniCAM scenarios generally represent the land as a growing carbon sink, with the exception of the Level 1 MiniCAM stabilization scenario, in which increased demand for land for biomass production leads to conversion and carbon loss. This effect is particularly strong prior to 2080 in the Level 1 MiniCAM stabilization scenario.

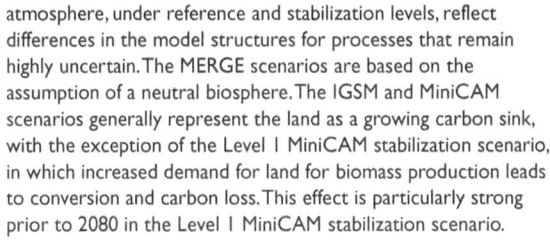

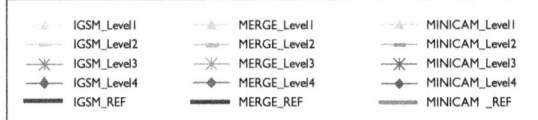

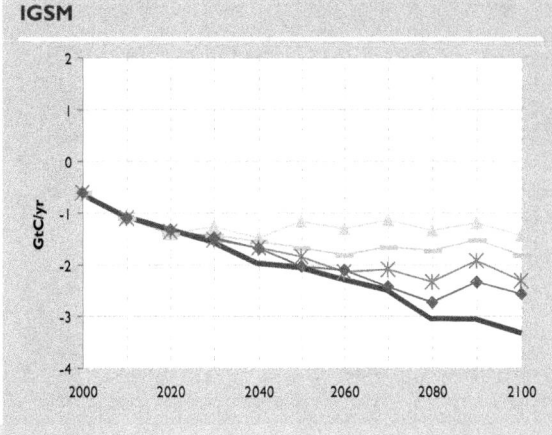

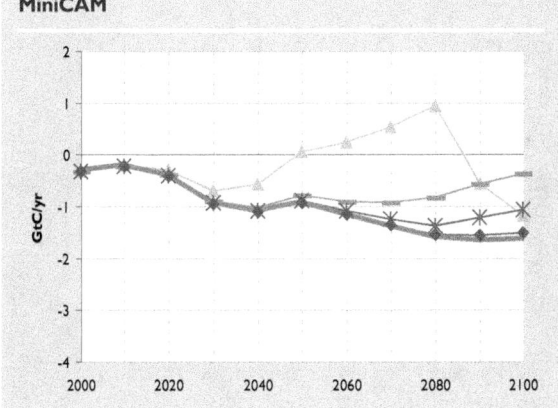

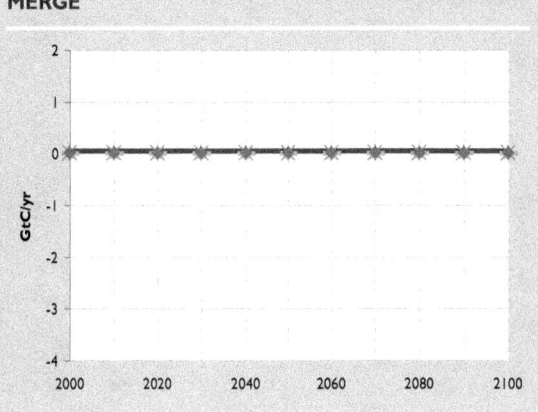

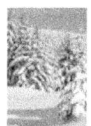

restrial biosphere in the IGSM and MiniCAM stabilization scenarios (Figure 4.19). The effect is larger and begins earlier the more stringent the stabilization level. For example, in the IGSM Level 4 scenario, the effect becomes substantial after 2070 and amounts to about 0.8 GtC/yr in 2100. The IGSM Level 1 scenario begins to depart markedly from the reference before 2050, and the departure from reference grows to approximately 2.0 GtC/yr by 2100. The effect of the diminished CO_2 fertilization effect is to require emissions mitigation in the energy-economy system to be larger by the amount of the difference between the reference aggregate net terrestrial CO_2 uptake and the uptake in the stabilization scenario. The MiniCAM stabilization scenarios exhibit similar carbon cycle behavior. The MERGE stabilization scenarios maintain the assumption of a neutral terrestrial biosphere as in the MERGE reference scenario.

The MiniCAM scenarios also include a second effect that results from the interaction between the energy system and emissions from changes in land use, such as converting previously unmanaged lands to bioenergy crop production. As in the IGSM scenarios, economic competition among alternative human activities, crops, pasture, managed forests, bioenergy crops, and unmanaged ecosystems determine land use. In the MiniCAM scenarios, this competition also determines land-use change emissions. One implication is increasing pressure to deforest under stabilization in order to clear space for biomass crops (Sands and Leimbach 2003). This effect is best exhibited in the Level 1 scenarios, in which the terrestrial biosphere becomes a net source of carbon rather than a sink from 2050 to past 2080. The effect subsides after 2080 because commercial biomass production ceases to expand beyond 2080, reducing any further pressure to deforest for biomass crops. Thus, terrestrial uptake in the MiniCAM scenarios is

Figure 4.20. Carbon Prices Across Stabilization Scenarios ($/tonne C, 2000$). In all the stabilization scenarios, the
carbon price rises, by design, over time until stabilization is
achieved (or the end-year 2100 is reached), and the prices are
higher the more stringent is the stabilization level. There are
substantial differences in carbon prices between MERGE and
MiniCAM stabilization scenarios, on the one hand, and the IGSM
stabilization scenarios on the other. Differences between the
models reflect differences in the emissions reductions necessary
for stabilization and differences in the technologies that might
facilitate carbon emissions reductions, particularly in the second
half of the century.

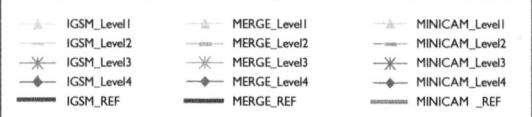

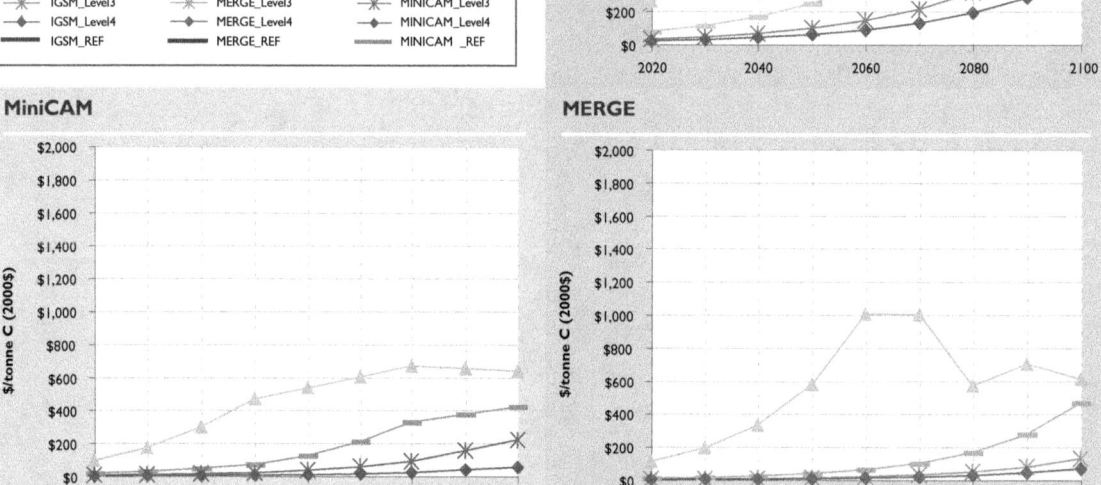

Table 4.5. Carbon Prices in 2020, 2030, 2050, and 2100 for Each Stabilization Scenario and Model.

Stabilization Level	2020 ($/tonne C)			2030 ($/tonne C)		
	IGSM	MERGE	MiniCAM	IGSM	MERGE	MiniCAM
Level 4	$18	$1	$1	$26	$2	$2
Level 3	$30	$2	$4	$44	$4	$7
Level 2	$75	$8	$15	$112	$13	$26
Level 1	$259	$110	$93	$384	$191	$170

Stabilization Level	2050 ($/tonne C)			2100 ($/tonne C)		
	IGSM	MERGE	MiniCAM	IGSM	MERGE	MiniCAM
Level 4	$58	$6	$5	$415	$67	$54
Level 3	$97	$11	$19	$686	$127	$221
Level 2	$245	$36	$69	$1,743	$466	$420
Level 1	$842	$574	$466	$6,053	$609	$635

reduced because of the lower CO_2 fertilization
effects as in the IGSM scenarios, and it is also
reduced by any land use change emissions that
derive from the increasing demand for bioenergy crops.

The terrestrial emissions reported in Figure 4.19
for the MiniCAM scenarios assume a policy architecture that places a value on energy and industrial emissions as well as carbon in terrestrial
systems. Thus, there is an economic incentive
to maintain and/or expand stocks of terrestrial

carbon as well as an incentive to bring more land under cultivation to grow bioenergy crops. Pricing terrestrial carbon exerts an important counter-pressure to deforestation and other land-use changes that generate increased emissions. To illustrate this effect, sensitivity cases were run by the MiniCAM modeling group in which no price was applied to terrestrial carbon emissions. These sensitivity analyses showed increased levels of land-use change emissions when terrestrial carbon was not valued, particularly at the more stringent stabilization levels, and the potential for a vicious cycle to emerge. Efforts to reduce emissions in the energy sector create an incentive to expand bioenergy production without a counter incentive to maintain carbon in terrestrial stocks. The resultant deforestation increases terrestrial CO_2 emissions, requiring even greater reductions in fossil fuel CO_2 emissions, even higher prices on fossil fuel carbon, and further increases in the demand for bioenergy, leading, in turn, to additional deforestation. The net terrestrial emissions for the MiniCAM scenarios reported here avoid this vicious cycle because they include a policy architecture that places a value on terrestrial carbon.

Despite the significant differences in the treatment of terrestrial systems in the three models, it is interesting to recall from Figure 3.20 that the overall behavior of the three carbon-cycle models is similar.

ECONOMIC IMPLICATIONS OF STABILIZATION

The economic implications of stabilization include increases in the prices of fossil fuels and electricity, along with reductions in economic output. Substantial differences in GHG emissions prices and associated economic costs arise among the modeling groups for each stabilization level. Among the most important factors influencing the variation in economic costs are: (1) differences in assumptions – such as those regarding economic growth over the century, the behavior of the oceans and terrestrial biosphere in taking up CO_2, and opportunities for reduction in non-CO_2 GHG emissions – that determine the amount that CO_2 emissions that must be reduced to meet the radiative forc-

ing stabilization levels; and (2) differences in assumptions about technologies, particularly in the second half of the century, to shift final demand to low-CO_2 sources such as biofuels and low-carbon electricity or hydrogen in transportation, industrial, and buildings end uses. Although differences in technology do not strongly emerge until the second half of the century, they cast a shadow over the full century because of the manner in which all three the modeling groups allocated carbon emissions reductions over time.

In most scenarios, carbon prices depress demand for fossil fuels and therefore their producer prices. Electricity producer prices generally increase because of increasing demand for electricity along with substitution to higher cost, lower emitting electricity production technologies. Consumer prices for all fuels (fuel price plus the carbon price for emitted carbon plus any added cost of capturing and storing carbon) are generally higher under the stabilization scenarios due to carbon price. The approaches to Non-CO_2 GHG prices differs among the modeling groups, reflecting differing approaches to the tradeoffs between reductions in the emissions of these GHGs and reductions in CO_2 emissions.

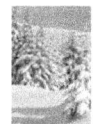

Stabilization and Carbon Prices

As discussed earlier, all of the modeling groups implemented prices or constraints that provide economic incentives to reduce GHG emissions. The instruments used to reduce CO_2 emissions in the models can be interpreted as the carbon price that would be consistent with either a universal cap-and-trade system or a harmonized carbon tax.

Across models, the more stringent stabilization levels require higher carbon prices because they require larger emissions reductions (Figure 4.20 and Table 4.5). Stabilization becomes increasingly difficult at the more stringent stabilization levels as can be seen in the difference in carbon prices between Level 2 and Level 1 as compared to that between Level 3 and Level 4. (Note that $100/tonne C is equivalent to $27/tonne CO_2. See Box 3.2 for more on converting between units of carbon and units of CO_c.)

Table 4.6. Cumulative Emissions Reductions Across Scenarios (GtC through 2100)

	IGSM	MERGE	MiniCAM
Level 4	472	112	97
Level 3	674	258	267
Level 2	932	520	541
Level 1	1172	899	934

Across models, the carbon prices rise exponentially throughout the century (in the IGSM scenarios) or until stabilization is reached (in the MERGE and MiniCAM scenarios). This similarity in the qualitative structure of the carbon price paths reflects the similarity in the approach that the modeling groups took to allocate emissions reductions over time, or *when* flexibility, as discussed in Section 4.2. This approach to *when* flexibility, with a carbon price that rises over time, tends to minimize the present discounted cost of emissions mitigation over the whole century. It also has the effect of linking future carbon prices to near-term carbon prices in a predictable way. Thus, when there are differences in technology assumptions that mostly appear in the second half of the century or

in reference emissions that occur mostly in the middle of the century, the assumption imposed on the price path means that the burden of emissions reduction is spread over the entire century. In this way, forces that do not emerge until mid-century or beyond cast a shadow onto the present.

At every stabilization level, there is variation in the carbon prices among the models. For example, the carbon price in 2100 exceeds $1700/tonne C in the IGSM Level 2 scenario while the carbon prices in the MERGE and MiniCAM Level 2 scenarios are $420 to $460/tonne C. The ratio among the models of carbon prices for other stabilization levels follows the same pattern. The range of carbon prices shown in these scenarios is consistent with other studies in the literature (IPCC 2001).

The carbon prices in the scenarios in this study are the result of a complex interplay of differing structural characteristics of the participating models and variation in key parameter values. Nonetheless major differences among carbon prices can be attributed to two influences: (1) the amount that emissions must be reduced to

Figure 4.21. Relationship Between Carbon Price and Percentage Emissions Reductions in 2050 and 2100. The relationship between carbon price and percentage reductions in carbon emissions is similar among the models in 2050. In 2100, a given percentage emissions reduction is generally more expensive in the IGSM stabilization scenarios than in the MERGE and MiniCAM stabilization scenarios. The difference in 2100 is due, in large part, to different assumptions regarding the technologies available to facilitate emissions reductions in the second half of the century, with IGSM scenarios assuming relatively fewer or more costly options than the scenarios from the other two modeling groups. *[Note. CO$_2$ emissions vary across the reference scenarios from the three modeling groups, so that similar percentage reductions, as shown in this figure, imply differing levels of total emissions reduction.]*

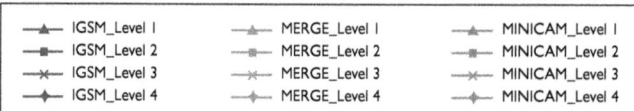

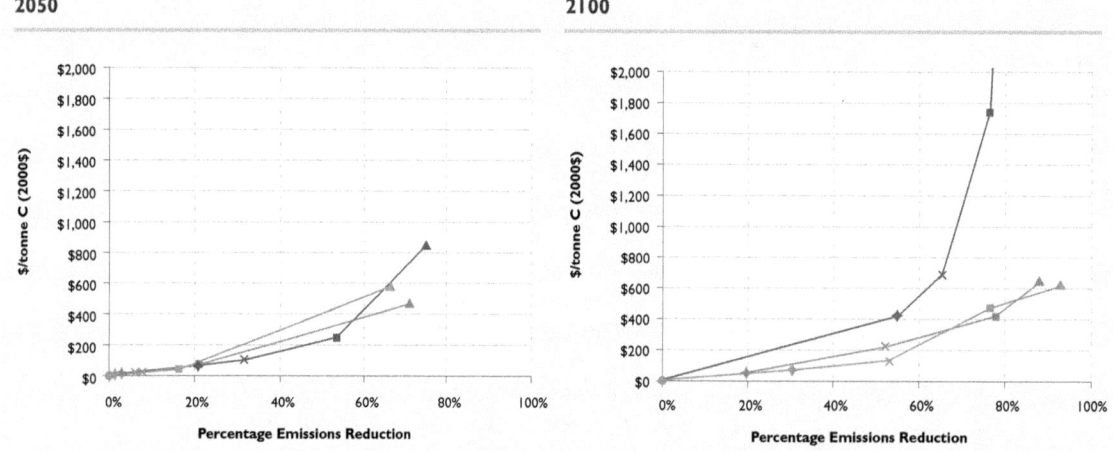

Figure 4.22. Percentage of Global Electricity Production from Low-or Zero-Emissions Technologies Across Scenarios (percent). All three modeling groups assumed sufficient technological options to allow for substantial reductions in carbon emissions from electricity production. Options include fossil power plants with CCS, nuclear power, and renewable energy such as hydroelectric power, wind power, and solar power. In all of the Level 1 scenarios, the electricity sector is almost fully decarbonized by the end of the century.

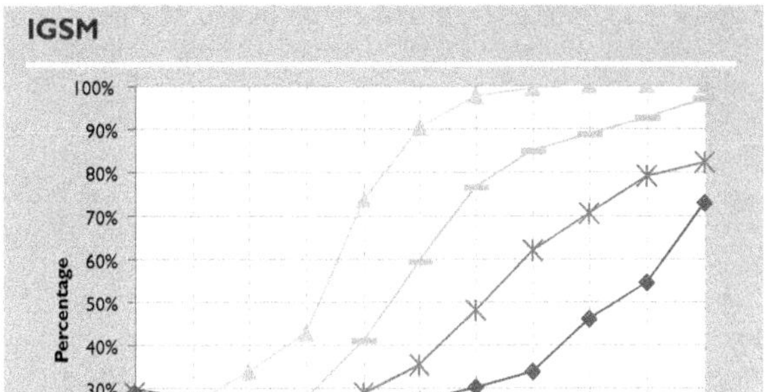

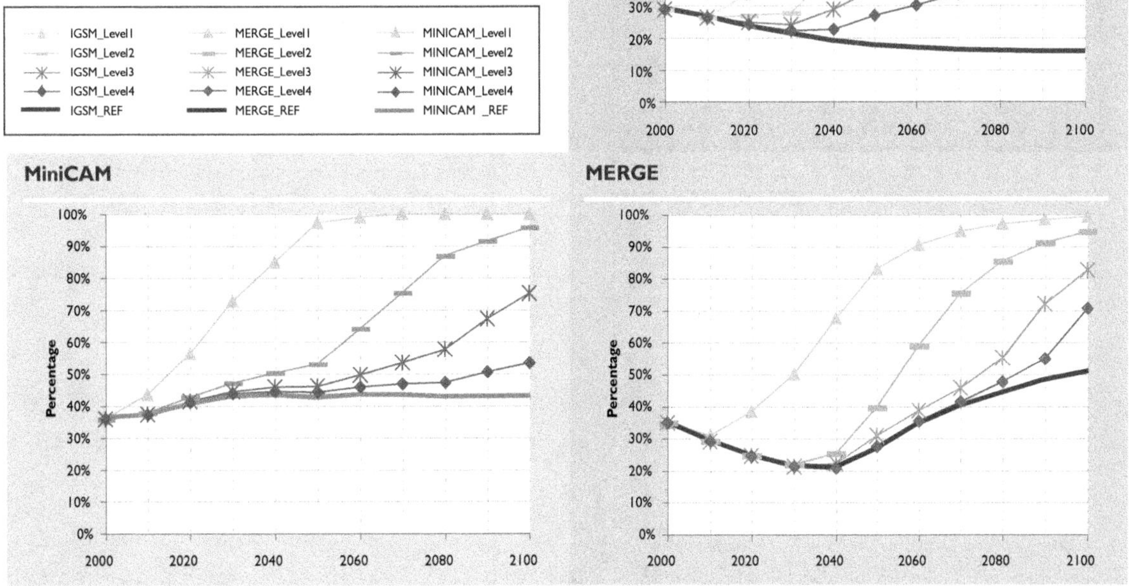

IGSM

IGSM_Level1	MERGE_Level1	MINICAM_Level1
IGSM_Level2	MERGE_Level2	MINICAM_Level2
IGSM_Level3	MERGE_Level3	MINICAM_Level3
IGSM_Level4	MERGE_Level4	MINICAM_Level4
IGSM_REF	MERGE_REF	MINICAM_REF

MiniCAM

MERGE

achieve an emissions path to stabilization, and (2) the technologies that are available to facilitate these changes in the economy.

On the first point, Table 4.6 shows the cumulative CO_2 emissions reductions required over the century across all four stabilization scenarios from each modeling group. Differences in total reductions come principally from three aspects of model behavior and assumptions: differences in forces, such as economic growth, that determine emissions in the reference scenario (Tables 3.2 and 3.3, and Figure 3.2); the behavior of the ocean and terrestrial systems in taking up carbon (Figure 4.6 and Figure 4.19); and the technological options available for constraining the emissions of non-CO_2 GHGs (Figure 4.8 and Figure 4.9). At all stabilization levels, the IGSM stabilization scenarios require greater CO_2 emissions reductions than the MERGE or MiniCAM stabilization scenarios. Indeed, the emissions reductions in the IGSM Level 2 scenario are commensurate with those of the MERGE and MiniCAM Level 1 scenarios. All other things being equal, the greater the required

emissions reductions the higher will be the emissions prices required to meet each target.

The second factor, the modeling of technology, also contributes to the differences among costs. The aggregate effect of differing technological assumptions is illustrated in Figure 4.21, which shows the relationship between the carbon price and percentage emissions reductions in 2050 and 2100 across all four stabilization scenarios from each modeling group. Roughly speaking, these figures represent what economists refer to as the marginal abatement cost functions for these periods. They broadly capture the technological opportunities for emissions reductions represented in the models. The similarity between the marginal abatement cost functions in 2050 implies that the technological opportunities represented by the three modeling groups are similar in 2050. The implication is that if the three modeling groups were to determine the carbon price associated with, for example, a 50% reduction in emissions in 2050, the results would be similar.

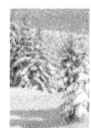

Figure 4.23. Percentage Reduction in World Primary Energy Consumption Across Scenarios (percent).
Differences in assumptions about technological opportunities result in different aggregate approaches to emissions reductions in the
stabilization scenarios from the three modeling groups. The IGSM
stabilization scenarios include greater reductions in primary
energy consumption than the MERGE and MiniCAM stabilization
scenarios because fewer technological opportunities, on both the
demand and supply side, are available for emissions reductions
through substitution to low or zero-carbon energy sources.
*[Note. Primary energy consumption from nuclear power and non-
biomass renewable electricity are accounted for at the average
efficiency of fossil-fired electric facilities, which vary over time and
across scenarios.]*

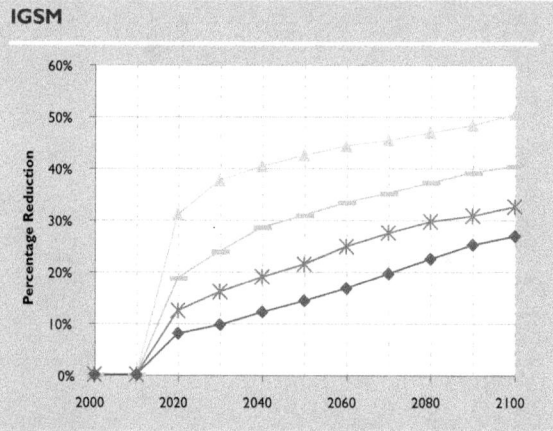

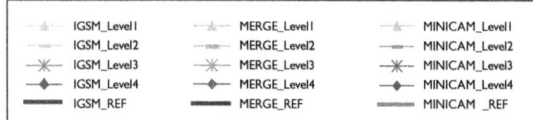

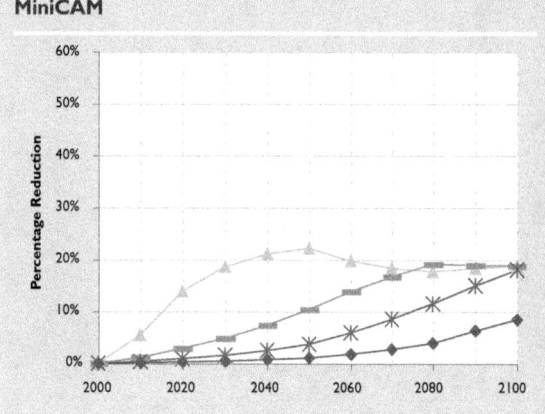

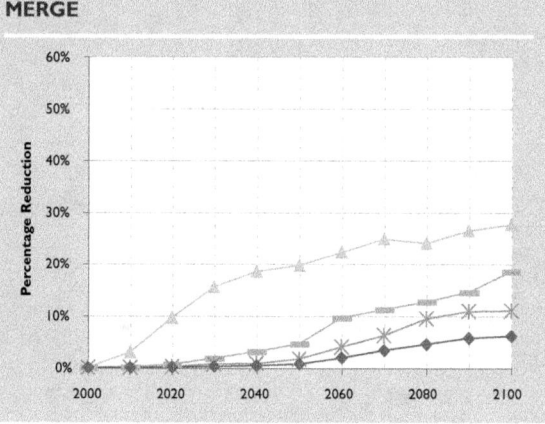

It is in the second half of the century that substantial differences in the marginal abatement cost functions emerge, particularly when the required abatement pushes towards and beyond 60% below the reference level as is the case in the Level 1 and Level 2 scenarios. There is no small set of technology assumptions used by the modeling groups that determines these differences. Among the modeling groups, assumptions about technology vary along a range of dimensions such as the rate of growth in labor productivity, the cost and performance of particular energy supply technologies, the productivity of agriculture and the associated costs of bioenergy, and the ability to substitute among various fuels and electricity in key demand sectors such as transportation. These assumptions are embodied not just in model parameters, but also, as discussed in Chapter 2, in the underlying mathematical structures of the models. As can be seen in Table 2.1, end-use technologies, are, in general, not represented explicitly. None of the participating models, for example, iden-

tify multiple steel production technologies or a wide range of vehicle options each with different energy using characteristics. Instead, energy demand responses are represented in relatively aggregate economic sectors (e.g., energy intensive industry or transportation). Other technologies, particularly in energy supply (e.g., CCS) are more likely to be identified specifically.

Three general characteristics of technology bear note with respect to the variation in carbon prices: (1) the availability of low- or zero-carbon electricity production technologies, (2) the supply of non-electric energy substitutes such as biofuels and hydrogen, and (3) the availability of technologies to facilitate substitution toward the use of electricity.

All three modeling groups assumed a variety of cost-effective technology options would be available to limit CO_2 emissions from electricity production. For example, the electric sector is almost fully de-carbonized by the end of the century in all three Level 1 scenarios (Figure 4.22). Electricity is produced with non-fossil technologies (nuclear or renewables) or fossil-fired power plants with CCS. Thus, although low carbon technologies in the electric sector do influence the carbon prices, it is forces outside of electricity production that drive costs at higher levels of abatement because options available to the electric sector can support its almost complete de-carbonization.

The second technology factor is the set of options available to substitute alternative, non-electric fuels for fossil energy in end-use sectors, most importantly in transportation. All three modeling groups assumed biofuels as a substitute for fossil fuels in non-electric applications. As discussed in Section 2 and Section 3, production of bioenergy crops must compete with other uses of agricultural lands in the IGSM and MiniCAM scenarios, which constrains total production of these substitutes. MERGE uses an aggregate parameterization to represent these same constraints. Even with these differing approaches, bioenergy production is similar across the stabilization scenarios. However, because of higher oil prices (Figure 3.7), the IGSM reference scenario includes substantial biofuels (Figure 4.10) so that expansion of biofuels is more limited in the IGSM stabilization scenarios.

In addition to biofuels, the MiniCAM and MERGE scenarios include other non-electric alternatives, and these become important for more stringent emissions reductions. The MERGE scenarios include a generic alternative fuel generated from renewable sources; which could be, for example, hydrogen from solar or wind power. In the MERGE Level 1 scenario, this alternative fuel provides roughly 80% as much non-electric energy as biofuels by 2100. The MiniCAM scenarios include hydrogen production using electricity, nuclear thermal dissociation, and fossil fuels with and without CCS. Though smaller than biofuels, the contribution of hydrogen rises to a little over 15% of global non-electric energy consumption in the Level 1 MiniCAM scenario. Without these additional options included in the MERGE and MiniCAM scenarios, the marginal cost of emissions reductions is higher in the IGSM scenarios, and more of the abatement is met through reductions in energy use (Figure 4.23).

Another factor influencing carbon prices at higher levels of emissions reduction is the ability to substitute to electricity in end-use sectors, through technologies such as heat pumps, electrically-generated process heat, or electric cars. Were all end uses to easily switch to electricity, then the availability of nearly carbon-free electricity production options in these scenarios would allow complete CO_2 emissions reduction at no more than the cost of these generation options. However, assumptions about technologies for electrification differ substantially among the modeling groups. The MERGE and MiniCAM modeling groups assumed greater opportunities for substitution to electricity than did the IGSM modeling group in the second half of the 21st century. As a result the electricity fraction of primary energy consumption is higher in the MERGE and MiniCAM scenarios in both the reference scenario and the stabilization scenarios, as shown in Figure 4.24. This means that low- or zero-carbon electricity production technologies can serve more effectively as a low-cost option for emissions reduction, reducing costs. In the IGSM scenarios, fuel demand for transportation, where electricity is not an option and for which biofuels supply is insufficient, continues to be a substantial source of emissions.

Although the main technological influences discussed above do not emerge for many decades, they influence carbon prices and economic costs from the outset because of the approach the modeling groups took to *when* flexibility, as discussed above. This dynamic view of the stabilization challenge reinforces the fact that actions taken today both influence and are influenced by the possible ways that the world might evolve in the future.

Finally, there are other structural differences among the approaches taken by the modeling groups that likely play a role in the variation in carbon prices. For example, MERGE is a forward-looking model and that behavior allows it to more fully optimize investments over time than the other two models, including investments in emissions reductions. Another difference is that the MiniCAM scenarios include CCS in cement production, which allows for cement emissions to be reduced to almost zero at more stringent stabilization levels. The IGSM scenarios include cement production within an aggregate sector so that mitigation options that

Figure 4.24. Ratio of Global Electricity Production to Primary Energy Consumption Across Scenarios. Efforts to constrain CO_2 emissions result in increased use of electricity as a fraction of total primary energy in the scenarios from all three modeling groups. This is because all three modeling groups assumed lower cost technology options for reductions in emissions from electricity production than for substitution away from fossil fuels in direct uses such as transportation. The MERGE and MiniCAM scenarios generally include greater electrification than the IGSM scenarios, with MiniCAM having the highest proportion of electricity to primary energy. Greater opportunities to electrify reduce the economic impacts of stabilization. *[Note. Primary energy consumption from nuclear power and non-biomass renewable electricity are accounted for at the average efficiency of fossil-fired electric facilities, which vary over time and across scenarios.]*

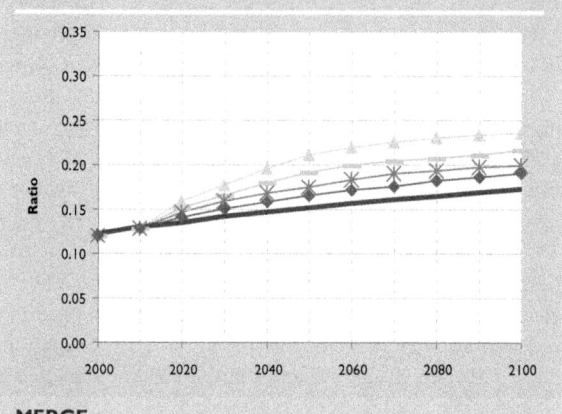

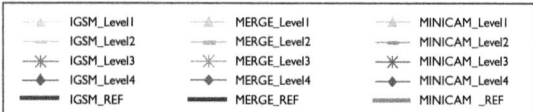

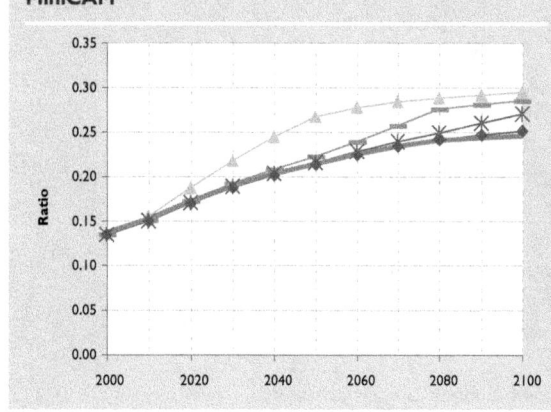

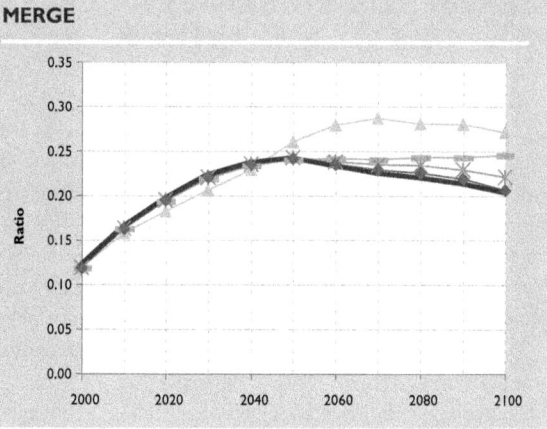

may be specific to this industry are not explicitly modeled. The MERGE scenarios explicitly include emissions from cement production, but do not include options for reducing these emissions. This omission puts more pressure on emissions reductions elsewhere in the IGSM and MERGE stabilization scenarios and would tend to raise carbon prices relative to the Mini-CAM scenarios. Finally, IGSM and MERGE explicitly track savings and investment, whereas MiniCAM does not. In IGSM and MERGE, investments in emissions reductions lower savings and investment in other sectors, affecting the scale of economic output in future periods, and this effect accumulates over time. The most direct effect of this dynamic is felt on economic output, and therefore stabilization costs (addressed later in this chapter), but it may also affect carbon prices through reductions in the scale of economic activity.

Stabilization and Non-CO_2 Greenhouse Gas Prices

Each of the three modeling groups employed a different approach to reductions in the emissions the non-CO_2 GHGs. After CO_2, CH_4 is the next largest component of radiative forcing in all three reference scenarios. Emissions of CH_4 vary among the reference scenarios. The IGSM reference scenario starts in the year 2000 at about 350 Mt/yr and rises to more than 700 Mt/yr (Figure 4.8), while the MERGE and MiniCAM scenarios begin with 300 Mt/yr in the year 2000. These are anthropogenic CH_4 emissions, and the differences reflect existing uncertainties in how much of total CH_4 emissions are from anthropogenic and natural sources. CH_4 emissions grow to almost 600 Mt/yr in the MERGE reference scenario. The MiniCAM reference scenario is characterized by a peak in CH_4 emission at less than 400 Mt/yr, followed by a decline to about 300 Mt/yr.

Each of the modeling groups took a different approach to setting a stabilization constraint on CH_4. The MiniCAM stabilization scenarios are based on constant GWP coefficients, so the price of CH_4 is simply the price of CO_2 multiplied by the GWP. This means that the price of CH_4 relative to the carbon price (the relative CH_4 price) is constant over time, as shown in Figure 4.25.

In contrast, MERGE determines the price of CH_4 to carbon through inter-temporal optimization. The relative price of CH_4 begins very low, although it is higher the more stringent the stabilization level. The relative price then rises at a roughly constant exponential rate of between 8% and 9% per year until stabilization is reached, at which point, the relative price remains approximately constant at around 10 times the carbon price. These characteristics of the CH_4 price and its relationship to the carbon price are the product of the inter-temporal optimization in which the long-term limit on radiative forcing is the only goal. Manne and Richels (2001) have shown that different patterns are possible if other formulations of the policy goal, such as limiting the rate of change of radiative forcing, are taken into account.

The IGSM stabilization scenarios are based on a third approach. CH_4 emissions are limited to a maximum value in each stabilization scenario: 425 Mt/yr at Level 4, 385 Mt/yr at Level 3, 350 Mt/yr at Level 2, and 305 Mt/yr at Level 1. As a consequence, the relative price of CH_4 initially grows from one-tenth to a maximum of between 3 and 14 between the years 2050 and 2080 and then declines thereafter. As previously discussed, this reflects an implicit assumption that a long-run requirement of stabilization means that eventually each substance must be (approximately) independently stabilized, and absent an explicit evaluation of damages of climate change, any time path of relative GHG prices cannot be determined.

As with CH_4, emissions of N_2O in the reference scenarios vary across the three modeling groups (Figure 4.9). The IGSM reference trajectory roughly doubles from approximately 11 Mt/yr to approximately 25 Mt/yr. In contrast, the MERGE and MiniCAM reference scenarios are roughly constant over time.

MERGE also sets the price of N_2O as part of the inter-temporal optimization process. The relative price trajectory for N_2O begins at roughly the level of the GWP-based relative price used in the MiniCAM stabilization scenarios and then rises, roughly linearly with time (Figure 4.25). The relative N_2O price approximately doubles in the MERGE Level 4 scenario, but is almost constant in the MERGE Level 1 scenario. Thus, in the Level 1 scenarios, the relative N_2O price path is virtually the same in the MERGE and MiniCAM scenarios.

In contrast, in the IGSM stabilization scenarios, stabilization sets a path to a predetermined N_2O concentration for each stabilization level, and the complexity of the price paths in Figure 4.25 shows the difficulty of stabilizing the atmospheric level of this GHG. Natural emissions of N_2O are calculated, which vary with the climate consequences of stabilization. The main anthropogenic source, agriculture, has a complicated relationship with the rest of the economy through the competition for land use.

The approaches employed by the three modeling groups do not necessarily lead to the stabilization of the concentrations of the non-CO_2 GHGs before the end of the twenty-first century, as concentrations are still rising slowly in some scenarios but below a long-term stabilized level (Figure 4.4 and Figure 4.5). How long-term stabilization was approached was independently developed by each modeling group.

Stabilization and Energy Prices

The carbon price drives a wedge between the producer prices of fuels and the costs to consumers. Table 4.7 provides an approximation of that of the relationship. A given carbon price has the largest impact on consumer cost of coal in percentage terms because the fuel price per unit of energy is low, and carbon emissions are relatively high per unit of energy. In comparison, natural gas prices were at historic highs in recent years and CO_2 emissions per unit of energy are lower than oil or coal. This means that the carbon price has a relatively smaller effect in comparison to the fuel price.

Stabilization scenarios tend to result in a lower producer price for oil (Figure 4.26). Stabilization

Figure 4.25. Relative Prices of CH$_4$ and N$_2$O to Carbon Price Across Scenarios (CH$_4$ in log scale). Differences in the prices of CH$_4$ and N$_2$O relative to the carbon price reflect different treatments of this tradeoff among the modeling groups, often referred to as *what* flexibility. In the MiniCAM stabilization scenarios, the tradeoff is based on the GWPs of the non-CO$_2$ GHGs, which are constants, leading to constant relative prices of the non-CO$_2$ GHGs. In the MERGE stabilization scenarios, relative prices are optimized with respect to meeting the long-run stabilization levels. In the IGSM stabilization scenarios, stabilization was forced for each GHG independently. Emissions were set so that concentrations of CH$_4$ would stabilize and allowed the CH$_4$ price path to be determined by changing opportunities for reducing emissions. Given N$_2$O emissions from agriculture, the relative price of N$_2$O is higher in the IGSM stabilization scenarios, in part because emissions were higher in the IGSM reference scenario than in the reference scenarios from the other two modeling groups. Lower emissions of N$_2$O for the MERGE and MiniCAM reference scenarios allowed the corresponding stabilization scenarios to achieve relatively low emissions at lower N$_2$O prices.

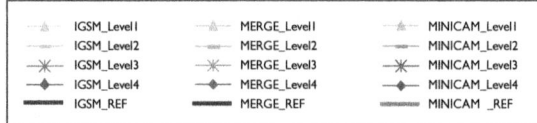

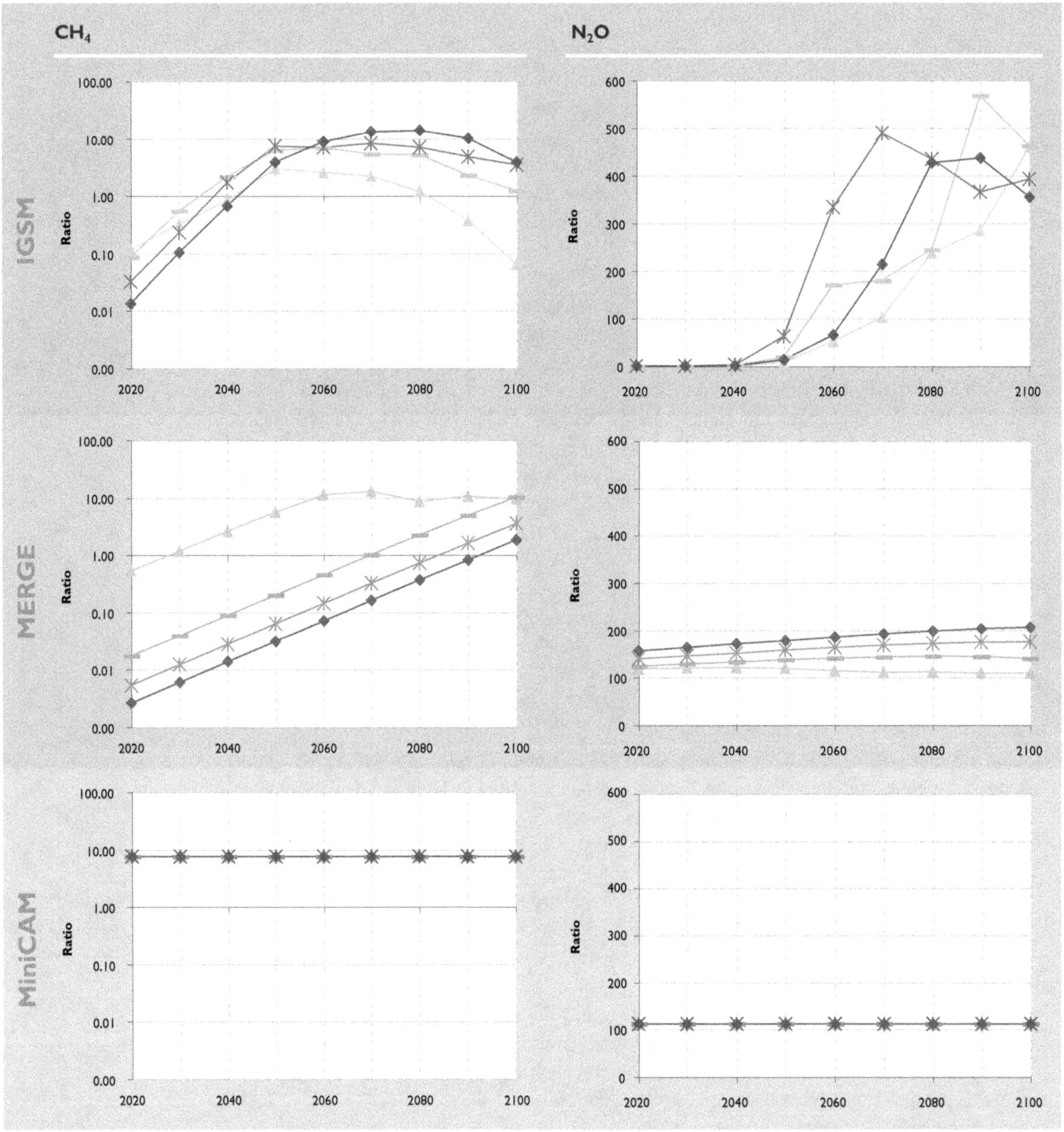

Fuel	Base Cost ($2005)	Added Cost ($)	Added Cost (%)
Crude Oil ($/bbl)	$60.0	$12.2	20%
Regular Gasoline ($/gal)	$2.39	$0.26	11%
Heating Oil ($/gal)	$2.34	$0.29	12%
Wellhead Natural Gas ($/tcf)	$10.17	$1.49	15%
Residential Natural Gas ($/tcf)	$15.30	$1.50	10%
Utility Coal ($/short ton)	$32.6	$55.3	170%
Electricity (c/kWh)	9.6¢	1.76¢	18%

Source: Bradley et al. (1991), updated with U.S. average prices for the 4th quarter of 2005 as reported by DOE (2006).

Table 4.7. Relationship Between a $100/tonne Carbon Price and Energy Prices. (In most cases, stabilization depresses producer prices and so the percentage rise in the fuel cost seen by consumers would be less than indicated here. The change in producer price is highly scenario and model dependent.)

at Level 4 has a relatively modest effect on the oil producer price, particularly prior to 2040; the effect is stronger the more stringent the stabilization level. Oil producer price reductions vary across the three models, ranging from the IGSM stabilization scenarios, which show the most pronounced effects, to the MERGE stabilization scenarios, which show a substantial effect only in the Level 1 scenario. The effect on world oil producer prices, in turn, depends on many factors, including how the supply of oil is characterized; the carbon price; and the availability of substi-

tute technologies for providing transportation liquids, such as bio-fuels or hydrogen.

Coal producer prices are similarly depressed in the IGSM and MiniCAM stabilization scenarios (Figure 4.27). The effect is mitigated by two features: (1) the assumed availability of CCS technology, which allows the continued large-scale use of coal in electricity production in the presence of a positive carbon price and (2) a coal supply schedule that is highly elastic. That is, demand for coal can exhibit large increases

Figure 4.26. World Oil Price Across Scenarios (Index, yr 2000 = 1). World oil prices (producer price) vary considerably across the reference scenarios. In all three models, stabilization tends to depress the producer prices of oil relative to the reference scenarios. [Note. Producer prices as defined here do not include additional costs associated with carbon emissions to the atmosphere through the combustion of fossil fuels, as shown in Table 4.7.]

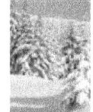

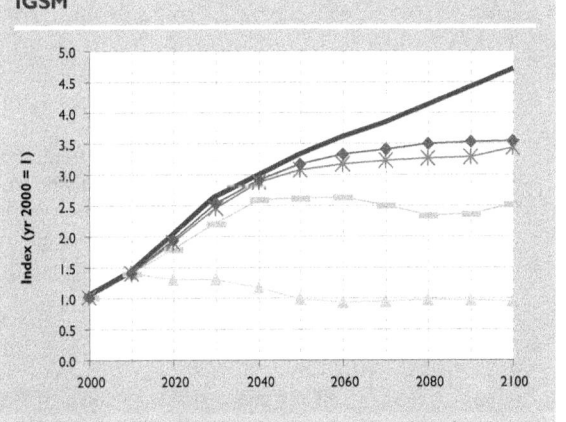

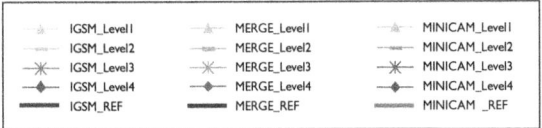

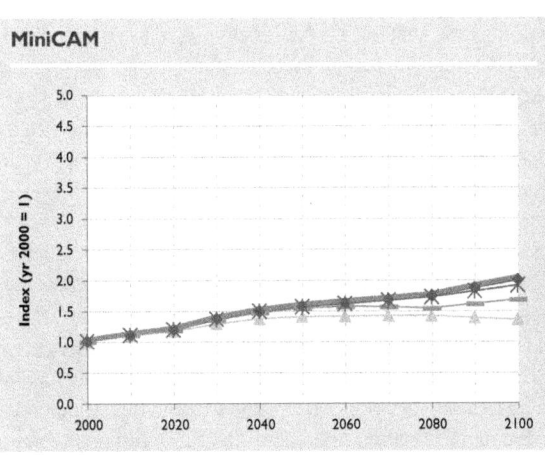

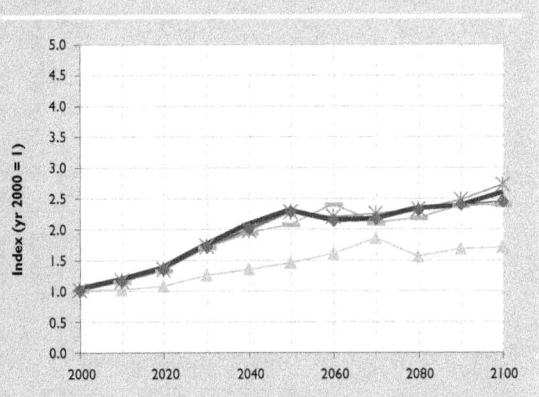

Figure 4.27. U.S. Mine-Mouth Coal Price Across Scenarios (Index, yr 2000 = 1). U.S. mine-mouth coal price varies acrosss the reference scenarios. In the IGSM and MiniCAM stabilization scenarios, stabilization depresses coal prices, whereas stabilization has no impact on coal prices in the MERGE stabilization scenarios, reflecting characterization of coal supply as an inexhaustible single grade such that there is no rent associated with the resource. Prices in the MERGE scenarios thus reflect the cost capital, labor, and other inputs that are little affected by the stabilization policy. *[Note. Producer prices as defined here do not include additional costs associated with carbon emissions to the atmosphere through the combustion of fossil fuels, as shown in Table 4.7.]*

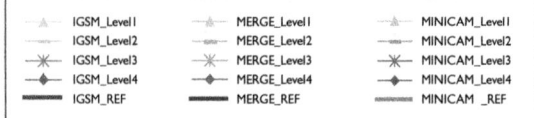

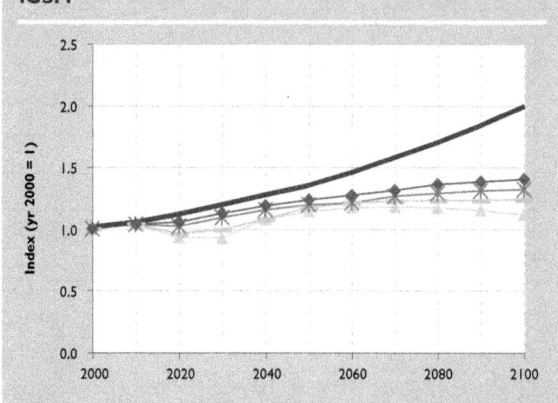

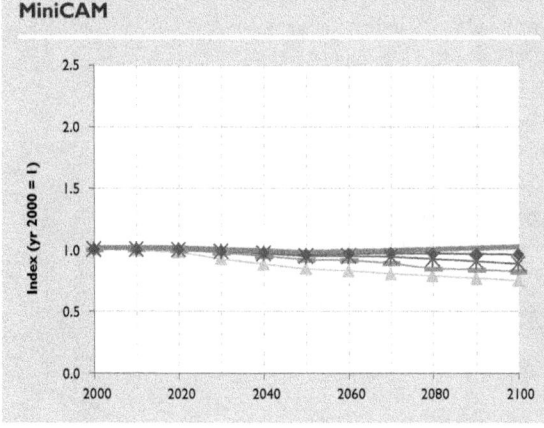

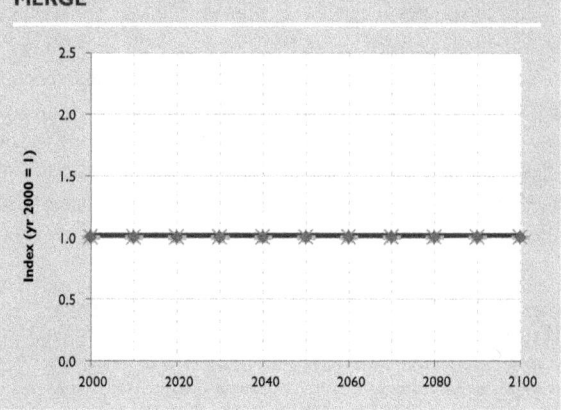

or decreases without much change in price. The high elasticity of supply in the MERGE scenarios leaves coal producer prices unchanged across the stabilization scenarios, whereas the MiniCAM and IGSM scenarios have lower supply price elasticities and, hence, greater producer price responses.

The impact on the natural gas producer price is more complex (Figure 4.28). Natural gas has roughly one-half the carbon-to-energy ratio of coal. Thus, emissions can be reduced without loss of available energy simply by substituting natural gas for coal or oil. As a consequence, two effects on the natural gas producer price work in opposite directions. With a postive carbon price, natural gas tends to substitute for other fossil fuels, increasing its demand. But a positive carbon price also means that a low- or zero-carbon substitutes, such as electricity, bioenergy, or energy-efficiency technologies, will tend to displace natural gas from markets, as happens for the more carbon-intensive fuels.

Thus, depending on the strength of these two effects, the producer price of natural gas can either rise or fall.

The natural gas producer price is most affected in the IGSM stabilization scenarios, reflecting the greater substitution of natural gas for coal in IGSM Level 2, 3, and 4 stabilization scenarios. In the IGSM Level 1 stabilization scenario, natural gas consumption is reduced over the entire period. On balance, the natural gas producer price is less affected by stabilization in the MERGE and MiniCAM scenarios in which the substitution and conservation effects are roughly offsetting.

Although the price that oil and coal producers receive tends to be either stable or depressed, that is not the full cost of using the fuel. Users, such as households or industrial fuel users, pay the market price plus the value of the carbon emissions associated with the fuel, which is the carbon price times the fuel's carbon-to-energy

Figure 4.28. U.S. Natural Gas Producers' Price Across Scenarios (Index, yr 2000 = 1). U.S. natural gas producers' prices vary among the reference scenarios. In the MiniCAM and MERGE stabilization scenarios, stabilization has little effect on the natural gas price. Stabilization at Levels 2, 3, and 4 increases the price of natural gas in the IGSM stabilization scenarios because of substitution toward natural gas and away from coal and oil. Natural gas prices fall relative to reference scenario in the IGSM Level 1 stabilization scenario because natural gas demand is depressed from the tight carbon constraint. *[Note. Producer prices as defined here do not include additional costs associated with carbon emissions to the atmosphere through the combustion of fossil fuels, as shown in Table 4.7.]*

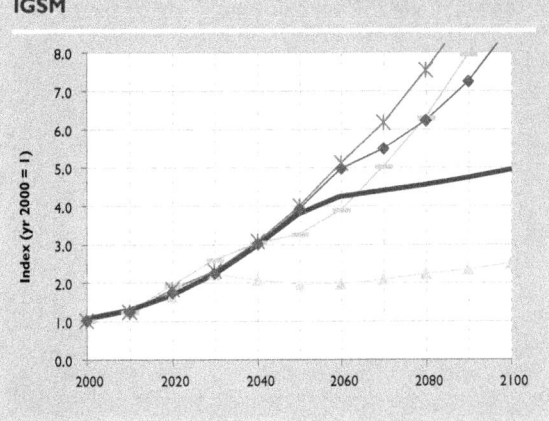

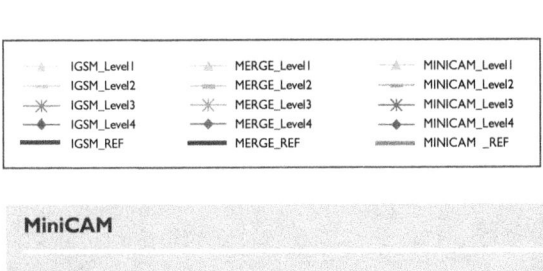

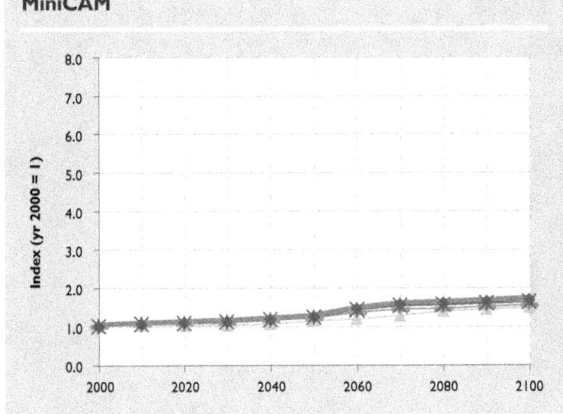

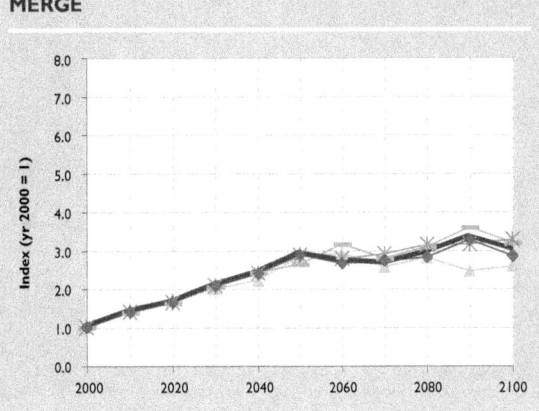

ratio. If they employ CCS, the carbon emissions are lower, but they face the added cost of CCS. Any additional carbon cost will be reflected in the users' fuel price if the carbon taxes, or required permits in a cap-and-trade system, are placed upstream with fuel producers. On the other hand, the actual fuel price impact they see may be similar to the producer price impact if carbon is regulated downstream where the fuel is consumed. In this case, users would be able to buy fuel relatively inexpensively, but would pay a separate large price for necessary carbon charges associated with emissions.

The effect on the price of electricity is another unambiguous result (Figure 4.29). Because electricity producers are fossil fuel consumers, the price of electricity contains the implicit carbon price in the fuels used for generation. All of the scenarios exhibit upward pressure on electricity prices, and the more stringent the stabilization level, the greater the upward pressure. The pressure is limited by the fact that there are

many options available to electricity producers to lower emissions. These options include, for example, the substitution of natural gas for coal; the use of CCS; the expanded use of nuclear power; the use of bioenergy; and the expanded use of wind, hydro, and other renewable energy sources.

The Total Cost of Stabilization

Assessing the macroeconomic cost of stabilization is not a simple task either conceptually or computationally. From an economic perspective, cost is the value of the loss in welfare associated with pursuing stabilization or equivalently, the value of activities that society will not be able to undertake as a consequence of pursuing stabilization. Although the concept is easy enough to articulate, defining an unambiguous measure is problematic. Any measure of cost is a more or less satisfactory compromise.

The task is further complicated by the need to aggregate the welfare of individuals who have

Figure 4.29. U.S. Electricity Producer Price Across Scenarios (Index, yr 2000 = 1). U.S. electricity prices in the reference scenarios range from little change over the century in the MiniCAM reference scenario to about a 50% increase from present levels in the IGSM reference scenario. Under stabilization, producer prices are affected by increasing use of more expensive low- or zero-emissions electricity technologies, including fossil electricity with CCS, nuclear power, and non-biomass renewables such as solar and wind power. Across the scenarios, rising fossil fuel prices are partially offset by increasing efficiency of fossil electric facilities. *[Note. Producer prices as defined here do not include additional costs associated with carbon emissions to the atmosphere through the combustion of fossil fuels, as shown in Table 4.7.]*

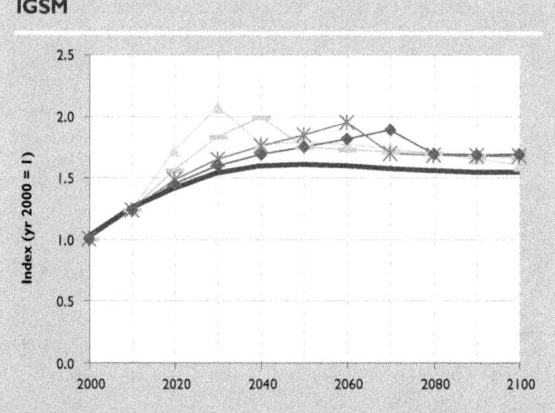

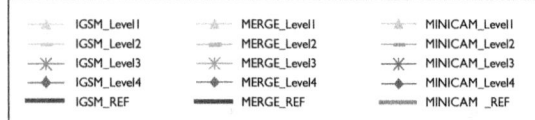

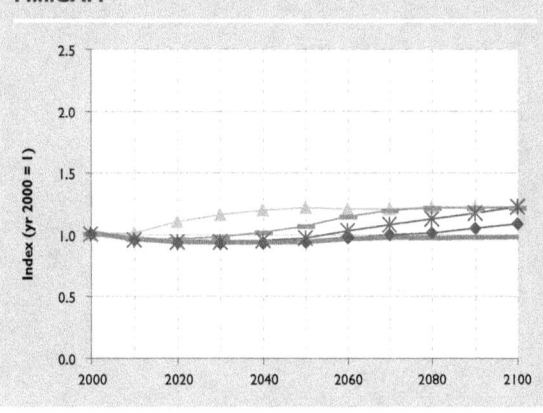

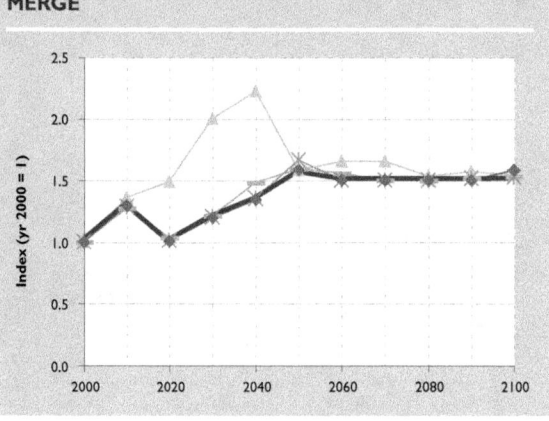

not yet been born and who may or may not share present preferences. Even if these problems were not difficult enough, economies can hardly be thought to currently be at a maximum of potential welfare. Preexisting market distortions impose costs on the economy, and mitigation actions may interact with them so as to reduce or exacerbate their effects. Any measure of global cost also runs into the problem of international purchasing power comparisons discussed in Chapter 3. Finally, climate change is only one of many public goods, and measures to address other public goods (like urban air quality) can either increase or decrease cost. To create a metric that is consistent and comparable across the three modeling platforms used in this study, all of these issues would have to be addressed in some way.

Beyond conceptual measurement issues, any metric including gross domestic product, depends on features of the scenario such as the assumed participation by countries of the world,

the terms of the emissions limitation regime, assumed efficiencies of markets, and technology availability – the latter including energy technologies, non-CO_2 GHG technologies, and related activities in non-energy sectors (e.g., crop productivity that strongly influences the availability and cost of producing commercial biomass energy). In almost every instance, scenarios of the type explored in this research employ more or less idealized representations of economic structure, political decision, and policy implementation (i.e., conditions that likely do not accurately reflect the real world, and these simplifications tend to lead to lower mitigation costs).

Finally, assessing welfare effects would require explicit consideration of how the burden of emissions reduction is shared among countries and the welfare consequences of income effects on poorer versus wealthier societies. Of course, if the world were to discover and deploy lower cost technology options than those assumed

Figure 4.30. Percentage Reduction in Gross World Product in the Stabilization Scenarios (percentage). Stabilization imposes costs on the economy, and stated in terms of gross world product, the costs rise over time as ever more stringent emissions restrictions are required. The more stringent the stabilization level, the higher the cost. The variation in costs among the models reflects differences in the emissions reductions necessary for stabilization and differences in the technologies that might facilitate carbon emissions reductions, particularly in the second half of the century.

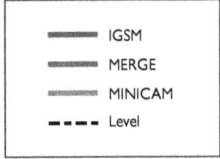

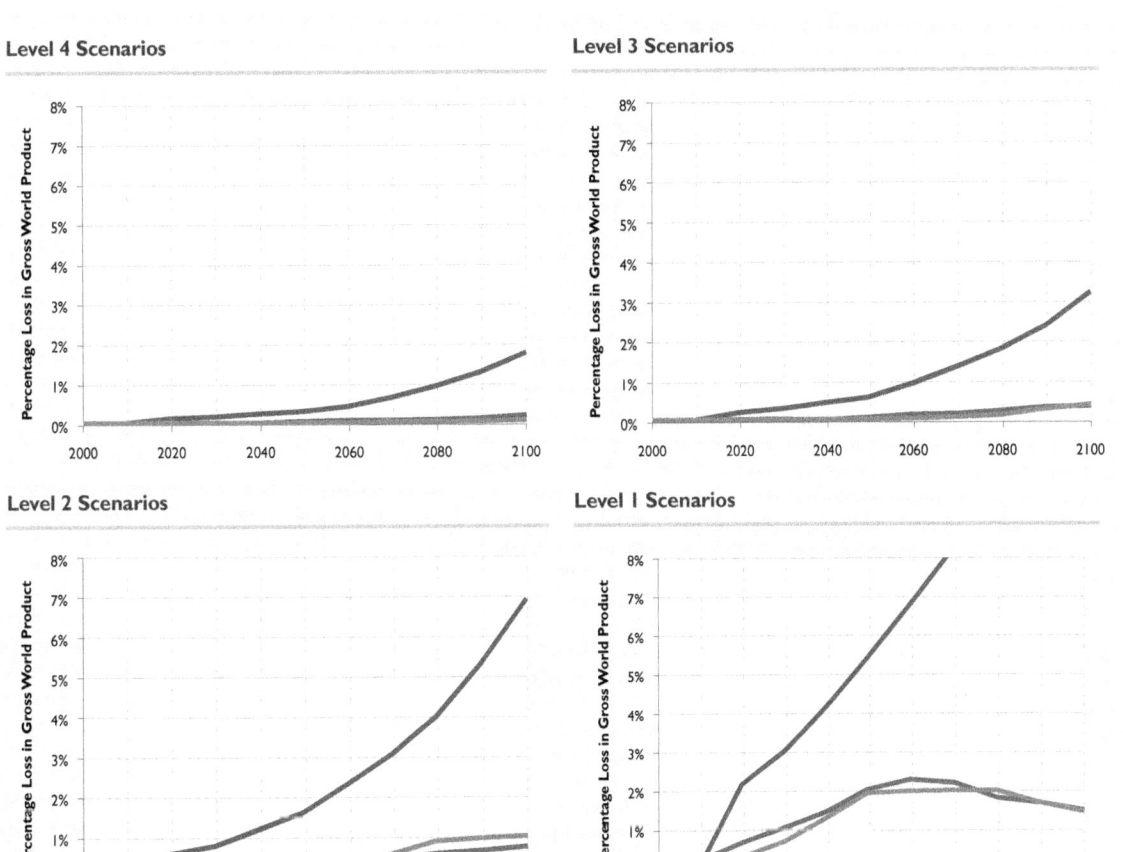

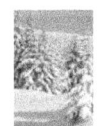

here, these costs could be lower. On the other hand, if society does not deliver the cost and performance for the technologies assumed in these scenarios, costs could be higher.

While all of the above considerations have not been extensively investigated in the literature, the implications of less-than-ideal implementation have been investigated, and these analyses show that it could increase the costs substantially. Richels et al. (1996) showed that for a simple policy regime, eliminating international *where* and *when* flexibility, while assuming perfect *where* flexibility within countries, could potentially raise costs by an order of magnitude compared to a policy that employed *where* and

when flexibility in all mitigation activities. Richels and Edmonds (1995) showed that stabilizing CO_2 emissions could be twice as expensive as stabilizing CO_2 concentrations and leave society with higher CO_2 concentrations. Babiker et al. (2000) similarly showed that limits on *where* flexibility within countries can substantially increase costs – although employing *where* flexibility also can increase costs in the context of tax distortions (Babiker et al. 2003a, Babiker et al. 2003b, Babiker et al. 2004, Paltsev et al. 2005).

Figure 4.30 reports the change of gross world product in the stabilization scenarios during the twenty-first century in the year in which it oc-

Table 4.8. Percentage Reduction in Gross World Product in the Stabilization Scenarios.

Level 1

	2020	2040	2060	2080	2100
IGSM	2.1%	4.1%	6.7%	10.1%	16.1%
MERGE	0.6%	1.4%	2.2%	1.8%	1.4%
MiniCAM	0.2%	1.2%	1.9%	1.9%	1.4%

Level 2

	2020	2040	2060	2080	2100
IGSM	0.5%	1.2%	2.3%	3.9%	6.8%
MERGE	0.0%	0.0%	0.4%	0.5%	0.7%
MiniCAM	0.0%	0.1%	0.2%	0.8%	1.0%

Level 3

	2020	2040	2060	2080	2100
IGSM	0.2%	0.4%	0.9%	1.8%	3.1%
MERGE	0.0%	0.0%	0.1%	0.2%	0.3%
MiniCAM	0.0%	0.0%	0.0%	0.1%	0.4%

Level 4

	2020	2040	2060	2080	2100
IGSM	0.1%	0.2%	0.4%	0.9%	1.7%
MERGE	0.0%	0.0%	0.1%	0.1%	0.2%
MiniCAM	0.0%	0.0%	0.0%	0.0%	0.0%

curs aggregated using market exchange rates. This information is also displayed in Table 4.8. The use of market exchange rates is a convenient choice given the formulations of the models employed here, but as discussed above and in Chapter 3 the approach has limits (see the Box 3.1 in Chapter 3). Though change in gross world product is not the most intellectually satisfying measure, it serves as a common reference point.

The effects on gross world product are tightly linked to the carbon prices. Therefore effects on gross world product in the stabilization scenarios follow the same patterns and logic as the carbon prices, which are discussed in substantially greater detail in Section 4.6.1. As with the carbon price, costs rise with increasing stringency of the stabilization level. And, as with the carbon price, there is variation in costs of stabilization among the modeling groups. For example, gross world product in 2100 is reduced by 6.8% in the IGSM Level 2 scenario, while the reduction is less than 1% in the

MERGE and MiniCAM Level 2 scenarios. The ratio of stabilization costs among the models at other radiative forcing stabilization levels follows the same pattern.

The differences in stabilization costs among the models can largely be attributed the same influences discussed in Section 4.6.1: (1) the amount that emissions must be reduced to achieve an emissions path to stabilization, and (2) the technologies that are available to facilitate these changes in the economy. A number of additional, structural differences, such as treatment of capital investment, intertemporal model structure, and emissions reductions opportunities in cement production also lead to differences in prices and costs. As with emissions prices, although technology differences emerge primarily in the second half of the century, their influence felt throughout the century because of the common implementation of *when* flexibility in the policy design.

Expressed throughout the report is the view that the development of independent sets of scenarios using three different models helps to inform common understanding of the forces that shape opportunities to stabilize greenhouse gas concentrations. The differences discussed here demonstrate the fundamental importance of technology in facilitating stabilization – particularly the importance of future technology, even developments more than half a century in the future. The scenarios also suggest the particular importance of options that facilitate the production of alternative non-electric fuels and demand-side technologies that will allow the substitution of electricity for current applications of fossil fuels.

Findings, Uses,
and Future Directions

CHAPTER 5

INTRODUCTION

Scenarios based on formal, computer-based models, such as the scenarios developed in this research, can help illustrate how key drivers such as economic and population growth or policy options lead to particular levels of GHG emissions. An important benefit of models such as those used in this research is that they ensure basic accounting identities and consistent application of behavioral assumptions. However, model-based scenarios are only one approach to scenario development, and models designed for one set of purposes may not be the most appropriate for other applications. Thus, the scenarios developed here should be viewed as complementary to other ways of thinking about the future, such as formal uncertainty analyses, story lines, baselines for further model-based scenarios, and analyses using other types of models.

The users of emissions scenarios are many and diverse and include climate modelers and the science community; those involved in national public policy formulation; managers of Federal research programs; state and local government officials who face decisions that might be affected by climate change and mitigation measures; and individual firms, non-governmental organizations, and members of the public. Such a varied clientele implies an equally diverse set of possible needs, and no single scenario exercise can hope to satisfy all of them. Scenario analysis is most effective when its developers can work directly with users, and initial scenarios lead to further *what if* questions that can be answered with additional scenarios or by probing more deeply into particular issues. The Prospectus for this research did not, however, prescribe such an interactive approach with a focused set of users. Instead, it called for a set of scenarios that provide broad insights into the energy, economic, and emissions implications of stabilizing radiative forcing. For the issue of stabilization, these scenarios are an initial offering to potential user communities that, if successful, will generate further questions and more detailed analysis.

This research focuses on three sets of scenarios, each including a reference scenario and four scenarios in which the radiative forcing from a common suite of GHGs is stabilized at four alternative levels. The stabilization scenarios describe a range of possible long-term goals for global climate policy. The stabilization levels imply a range of policy efforts and levels of urgency, from relatively little deviation from reference scenarios over the course of the century to major deviations starting very soon. Although the Prospectus did not mandate a formal treatment of likelihood or uncertainty, such analysis could be a useful follow-on activity. Here, however, the range of outcomes from the different modeling groups helps to illustrate, if incompletely, the range of possibilities.

For this research, a scenario is an illustration of future developments based on a model of the economy and the Earth system, applying a plausible set of model parameters and providing a basis for future work. None of the reference scenarios is a prediction or best-judgment forecast of the future, and none can be said to have the highest probability of being right. Nor does any single stabilization scenario provide the most correct picture of the changes to energy and other systems that would be required for stabilization. Instead, each scenario in this report is a thought experiment that helps illuminate the implications of different long-term policy goals.

OVERVIEW OF THE SCENARIOS

The scenarios are presented in text and figures in Chapters 3 and 4, and here a summary is provided of some of their key characteristics, some of the magnitudes involved, and the assumptions that lie behind them.

Reference Scenarios

The difficulty in achieving any specified radiative forcing stabilization level depends heavily on the emissions that would occur absent actions to address GHG emissions. In other words, the reference scenario strongly influences the stabilization scenarios. If the reference scenario has inexpensive fossil fuels and high economic growth, then larger changes to the energy sector and other parts of the economy may be required to stabilize the atmosphere. On the other hand, if the reference scenario shows lower growth and emissions, and perhaps increased exploitation of non-fossil sources even in the absence of climate policy, then the effort required to stabilize radiative forcing will not be as great.

Energy production, transformation, and consumption are central features in all of these scenarios, although non-CO_2 gases and changes in land use also make a significant contribution to aggregate GHG emissions. Demand for energy over the coming century will be driven by economic growth and will also be strongly influenced by the way that energy systems respond to depletion of resources, changes in prices, and improvements in technology. Demand for energy in developed countries remains strong in

all the scenarios and is even stronger in developing countries, where millions of people seek greater access to commercial energy. These developments strongly influence the emissions of GHGs, their disposition, and the resulting change in radiative forcing in the reference scenarios.

The three reference scenarios show the implications of this increasing demand and the improved access to energy. The variation between the reference scenarios reflects the differing assumptions used by the modeling groups.

- Global primary energy consumption rises substantially in all three reference scenarios, from about 400 EJ/yr in 2000 to between roughly 1275 EJ/yr and 1500 EJ/yr in 2100 (Figure ES.1). U.S. primary energy consumption also grows substantially, about 1¼ to 2½ times present levels by 2100. Primary energy consumption growth occurs despite continued improvements in the efficiency of energy use and energy production technologies. For example, the U.S. energy intensity – the ratio of primary energy consumption to economic output – declines 60% to 75% between 2000 and 2100 across the three reference scenarios.

- All three reference scenarios include a gradual reduction in the consumption of conventional oil resources. However, in all three, a range of alternative fossil-based resources, such as synthetic fuels from coal and unconventional oil resources (e.g., tar sands and oil shales), are available and become economically viable. Fossil fuels provided almost 90% of the global primary energy in 2000, and they remain the dominant energy source in the three reference scenarios throughout the twenty-first century, supplying 70% to 80% of total primary energy in 2100.

- Non-fossil fuel energy use also grows over the century in all three reference scenarios. Contributions to primary energy consumption in 2100 range from 250 EJ to 450 EJ – a range that at the hight end exceeds global primary energy consumption today. Despite this growth, these sources never supplant fossil fuels, although they provide an increasing share of the total, particularly in the second half of the century.

- Consistent with the characteristics of primary energy consumption, global and U.S. electricity production continues to rely on coal, although this contribution varies among the reference scenarios. The contribution of renewable and nuclear energy varies considerably in the different reference scenarios, depending on resource availability, technology, and non-climate policy considerations. For example, global nuclear power in the reference scenarios ranges from about 1½ times current levels (if non-climate concerns such as safety, waste, and proliferation constrain its growth as is the case in one reference scenario), to an expansion of almost an order of magnitude assuming relative economics as the only constraint.

- Oil and natural gas producer prices rise through the century relative to year 2000 levels, whereas coal and electricity prices remain relatively stable. It should be emphasized that the models used in this research were not designed to simulate short-term fuel-price spikes, such as those that occurred in the 1970s, early 1980s, and more recently in 2005. Thus, price trends in the scenarios should be interpreted as multi-year averages.

- As a combined result of all these influences, CO_2 emissions from fossil fuel combustion and industrial processes in the reference scenarios increase from approximately 7 GtC/yr in 2000 to between 22.5 GtC/yr and 24.0 GtC/yr in 2100; that is, to roughly 3 to 3½ times current levels.

The non-CO_2 GHGs, CH_4, N_2O, SF_6, PFCs, and HFCs, are emitted from various sources including agriculture, waste management, biomass burning, fossil fuel production and consumption, and a number of industrial activities.

- Future global anthropogenic emissions of CH_4 and N_2O vary widely among the reference scenarios, ranging from flat or declining emissions to increases of 2 to 2½ times present levels. These differences reflect alternative assumptions about technological opportunities and about whether current emissions rates will be reduced significantly for non-climate reasons, such as air pollution

control and/or higher natural gas prices that would further stimulate the capture of CH_4 emissions for its fuel value.

Increases in emissions from the global energy system and other human activities lead to higher atmospheric GHG concentrations and radiative forcing. This increase is moderated by natural biogeochemical removal processes.

- The oceans are a major sink for CO_2, and the rate at which they take up CO_2 generally increases in the reference scenarios as concentrations rise early in the century. However, processes in the ocean can slow this rate of increase at high concentrations late in the century. Ocean uptake in the three reference scenarios is in the range of 2 GtC/yr in 2000, rising to about 5 GtC/yr to 11 GtC/yr by 2100. The three ocean models behave more similarly in the stabilization scenarios; for example, the difference in ocean uptake among the models at the most stringent stabilization levels is less than 1 GtC/yr in 2100.

- Two of the three participating models include sub-models of the exchange of CO_2 with the terrestrial biosphere, including the net uptake by plants and soils and the emissions from deforestation. In the reference scenarios from these modeling groups, the terrestrial biosphere acts as a small annual net sink (less than 1 GtC/yr of carbon) in 2000, increasing to an annual net sink of roughly 2 GtC/yr to 3 GtC/yr by the end of the century. The third modeling group assumed a zero net exchange. Changes in emissions from terrestrial systems over time in the reference scenarios reflect assumptions about human activity (including a decline in deforestation) as well as increased CO_2 uptake by vegetation as a result of the positive effect of CO_2 on plant growth. There remains substantial uncertainty about this carbon fertilization effect and its evolution under a changing climate.

- As a result of the various influences, GHG concentrations rise substantially over the century in the reference scenarios. By 2100, CO_2 concentrations range from about 700 ppmv to 900 ppmv, up from 365 ppmv in 1998. CH_4 concentrations in 2100 range from 2000 ppbv to 4000 ppbv, up from 1745 ppbv in 1998, and N_2O concentrations in 2100 range from about 375 ppbv to 500 ppbv, up from 314 ppbv in 1998.

- As a result, radiative forcing in 2100 ranges from 6.4 W/m² to 8.6 W/m² from preindustrial, up from a little over 2 W/m² today. The non-CO_2 GHGs account for about 20% to 25% of radiative forcing at the end of the century.

Stabilization Scenarios

Important assumptions underlying the stabilization scenarios include the flexibility that exists in a policy design, as represented by the modeling groups, to seek out least cost options for emissions control regardless of where they occur, what substances are controlled, or when they occur. This set of conditions is referred to as *where*, *what*, and *when* flexibility. Equal marginal costs of abatement among regions across time (taking into account discount rates and the lifetimes of substances), and among substances (taking into account their relative warming potential and different lifetimes) will, under specified conditions, lead to least cost abatement. Each modeling group applied an economic instrument that priced GHGs in a manner consistent with the group's interpretation of *where*, *what*, and *when* flexibility. The economic characteristics of the scenarios thus assume a policy designed with the intent of achieving the required reductions in GHG emissions in a least-cost way. Key implications of these assumptions are that: (1) all nations proceed together in restricting GHG emissions from 2012 and continue together throughout the century, and that the same marginal cost is applied across sectors (*where* flexibility); (2) the marginal cost of abatement rises over time in these three sets of scenarios based on each modeling group's interpretation of *when* flexibility, with the effect of linking emissions mitigation efforts over the time horizon of the scenarios; and (3) stabilization of radiative forcing is achieved by combining control of all GHGs, with differences in how modeling groups compared them and assessed the implications of this *what* flexibility.

Although these assumptions are convenient for analytical purposes, to gain an impression of the

implications of stabilization, they are idealized versions of possible outcomes. For the abatement costs in these scenarios to be representative of actual abatement costs would require, among other things, that a negotiated international agreement include these flexibility mechanisms. Failure in that regard could have a substantial effect on the difficulty of achieving any of the stabilization levels considered in this research. For example, a delay of many years in the participation of some large countries would require greater effort by the others, and policies that impose differential burdens on different sectors without mechanisms to allow for equalizing marginal costs across sectors can result in a many-fold increase in the cost of any environmental gain. Therefore, *it is important to view these scenarios as representing possible futures under specified conditions, not as forecasts of the most likely outcome within the national and international political system.* Further, none of the scenarios considered the extent to which variation from these least-cost rules might be improved upon given interactions with existing taxes, technology spillovers, or other non-market externalities.

If the developments in the three reference scenarios were to occur, concerted efforts to reduce GHG emissions would be required to stabilize radiative forcing at the levels considered in this research. Such limits would shape technology deployment throughout the century and have important economic consequences. The stabilization scenarios demonstrate that there is no single technology pathway consistent with a given level of radiative forcing. Furthermore, there are other possible pathways than those considered in this research.

- Stabilization efforts are made more challenging by the fact that ocean uptake of CO_2 declines as the stringency of the stabilization level increases, and, in the scenarios from two of the models, because CO_2 uptake in terrestrial systems also declines with the stringency of the stabilization level.

- Stabilization of radiative forcing at the levels examined in this research would require a substantially different energy system globally, and in the U.S., than what emerges in the reference scenarios. The degree and timing of change in the global energy system depends on the level at which radiative forcing is stabilized. The lower the radiative forcing stabilization level, the larger the scale of change in the global energy system relative to the reference scenario and the sooner those changes would need to occur.

- Across the stabilization scenarios, the energy system relies more heavily on non-fossil energy sources, such as nuclear, solar, wind, biomass, and other renewable energy forms, than in the associated reference scenarios. The stabilization scenarios differ in the degree to which these technologies are deployed, depending on assumptions about: technological improvements; the ability to overcome obstacles, such as intermittency in the case of solar and wind power, or safety, waste, and proliferation issues in the case of nuclear power; and the policy environment surrounding these technologies. Energy consumption, while still higher than today's levels, is lower in the stabilization scenarios than in the reference scenarios.

- CCS is widely deployed in the stabilization scenarios because each modeling group assumed that the technology can be successfully developed and that concerns about storing large amounts of carbon do not impede its expansion. Removal of this assumption would make the stabilization levels more difficult to achieve and would lead to greater demand for low-carbon sources such as renewable energy and nuclear power, to the extent that growth of these other sources is not otherwise constrained.

- Significant fossil fuel use continues across the stabilization scenarios, both because stabilization allows for some level of carbon emissions through 2100, depending on the stabilization level, and because of the presence of CCS technology in all the stabilization scenarios.

- Emissions of non-CO_2 GHGs, such as CH_4, N_2O, HFCs, PFCs, and SF_6, are all reduced in the stabilization scenarios.

- Increased use is made of biomass energy crops in all the stabilization scenarios, but their contribution is ultimately limited by competition with agriculture and forestry, and, in one participating model, by the associated impacts of biomass expansion on carbon emissions from changes in land use.

- The lower the radiative forcing stabilization level, the larger the scale of change in the global energy system relative to the reference scenario required over the coming century and the sooner those changes would need to occur.

- Across the stabilization scenarios, the scale of the emissions reductions required relative to the reference scenario increases over time, with the bulk of emissions reductions taking place in the second half of the century. But emissions reductions occur in the first half of the century in every stabilization scenario.

- The 2100 time horizon of this research limited examination of the ultimate stabilization requirements. Further reductions in CO_2 emissions after 2100 would be required in all of the stabilization scenarios, because stabilization of radiative forcing at any of the levels considered in this research requires human emissions of CO_2 in the long term to be essentially halted. Despite the fact that much of the carbon emissions will eventually make its way into oceans and terrestrial sinks, some will remain in the atmosphere for thousands of years. Only CCS can allow continued burning of fossil fuels. Higher radiative forcing limits can delay the point in time at which emissions must be reduced toward zero, but this requirement must ultimately be met.

Fuel sources and electricity generation technologies change substantially, both globally and in the U.S., in the stabilization scenarios compared to the reference scenarios. There are a variety of technological options in the electricity sector that reduce carbon emissions in these scenarios.

- Nuclear power, renewable energy, and CCS all play important roles in stabilization scenarios. The contribution of each varies, depending on assumptions about technological improvements, the ability to overcome obstacles such as intermittency of supply, and the policy environment surrounding them.

- By the end of the century, electricity produced by conventional fossil technology that freely emits CO_2 is reduced in the stabilization scenarios relative to reference scenarios. Electricity production from technologies that emit CO_2 varies substantially with the stabilization level; in the lowest stabilization level, electricity production from these technologies is reduced toward zero.

The economic effects of stabilization are substantial in many of the stabilization scenarios, although much of this cost is borne later in the century. As noted earlier, each of the modeling groups assumed that a global policy was implemented after 2012, with universal participation by the world's nations, and that the time path of reductions approximated a least-cost solution. These assumptions of *where*, *when*, and *what* flexibility lower the economic consequences of stabilization relative to what they might be with other implementation approaches.

- The stabilization scenarios follow a pattern where, in most scenarios, the carbon price rises steadily over time, providing an opportunity for the energy system to adjust gradually.

- Although the general shape of the carbon price trajectory over time is similar across the models, the carbon prices vary substantially across the models. For example, two of the scenarios have prices of $10 or below per tonne of carbon in 2020 for the less stringent scenarios, with their prices rising to roughly $100 per tonne in 2020 for the most stringent stabilization level. A third scenario shows higher initial carbon prices in 2020, ranging from around $20 for the least stringent stabilization level to over $250 for the most stringent stabilization level.

- Factors contributing to differences in carbon prices include (1) differences in assumptions – such as those regarding economic growth over the century, the behavior of the oceans and terrestrial biosphere in taking up CO_2, and opportunities for reduction in non-CO_2 GHG emissions – that determine the amount that CO_2 emissions must be reduced to meet the radiative forcing sta-

bilization levels; and (2) differences in assumptions about technologies, particularly in the second half of the century, to shift final demand to low-carbon sources such as biofuels and low-carbon electricity or hydrogen, in transportation, industrial, and buildings end uses. Differences among the scenarios reflect the uncertainty that attends the far future.

- Differences in non-CO_2 gases also contribute to differences in abatement costs. Scenarios that assume relatively better performance of non-CO_2 emissions mitigation require less CO_2 abatement and therefore less stringent changes in the energy system, to meet the same overall radiative forcing goal.

- These differences in carbon prices, along with other model features, lead to similar variation in the costs of stabilization. At the most stringent radiative forcing stabilization level, for example, gross world product (aggregating country figures using market exchange rates) is reduced in 2050 by around 1% in the scenarios from two of the modeling groups and approximately 5% in the scenario from the third, and in 2100 it is reduced by less than 2% in two of the scenarios and over 16% in the third.

- The assumption of *when* flexibility links elements of the stabilization scenarios through time. This in turn means that, in addition to near-term technology availability, differences in assumptions about technology in the post-2050 period are also reflected in near-term emissions reductions and GHG prices.

- In all of the stabilization scenarios, emissions reductions in electric power sector come at relatively lower prices than in other sectors (e.g., buildings, industry, and transport) so that the electricity sector is essentially decarbonized in the most stringent scenarios. At somewhat higher cost other sectors can respond to rising carbon prices by reducing demands for fossil fuels, applying CCS technologies where possible, and substituting low-carbon energy sources such as bioenergy and low-carbon electricity or hydrogen. The amount of electricity used per unit of total primary energy increases in all of the stabilization scenarios, but those scenarios with the highest relative use of electricity tend to exhibit lower stabilization costs in part because of the larger role of decarbonized power generation. Assumptions regarding costs and performance of technologies to facilitate these adjustments, particularly in the post-2050 period, play an important role in determining stabilization costs

- As noted earlier, the overall cost levels are strongly influenced by the idealized policy scenario that has all countries participating from the start, the assumption of *where* flexibility, an efficient pattern of emissions reductions over time, and integrated reductions in emissions of the different GHGs. Assumptions in which policies are implemented in a less efficient manner would lead to higher cost. Thus, these scenarios should not be interpreted as applying beyond the particular conditions assumed.

- GHG mitigation would also affect fuel prices. Generally, producer prices for fossil fuels fall as demand for them is depressed by the stabilization measures. Consumers of fossil fuels, on the other hand, pay for fuel plus a carbon price if the CO_2 emissions are freely released to the atmosphere. Therefore, consumer costs of energy rise with more stringent stabilization levels in these scenarios.

Achieving stabilization of atmospheric GHGs poses a substantial technological and policy challenge. It would require important transformations of the global energy system. The cost and feasibility of such a goal depends on the evolution of technology and its ability to overcome existing limits and barriers to adoption, and it depends on the efficiency and effectiveness of the policy instruments employed to achieve stabilization.

APPLICATION OF THE SCENARIOS IN FURTHER ANALYSIS

These scenarios, supported by the accompanying database described in the Appendix, can be used as the basis of further analysis. There are a variety of possible applications for these scenarios. For example, the scenarios could be used as the basis for analysis of the climate implications, and then follow-on studies of potential climate impacts. Such studies might begin with the radiative forcing levels of each scenario, with the individual GHG concentrations (applying separate radiation codes) or with the emissions (applying separate models of the carbon cycle and of the atmospheric chemistry of the non-CO_2 GHGs). Such applications could be made directly in climate models that do not incorporate a three-dimensional atmosphere and detailed biosphere model. For the larger models, some approximation would need to be imposed to allocate the short-lived gases by latitude or grid cell. Such an effort would need to include scenarios of the emissions (or concentrations) of the reflecting and absorbing aerosols. This could be achieved by the use of sub-models linked to scenario for energy use by fuel.

The scenarios could also be used as a point of departure for partial equilibrium analysis of technology development. Because these models compute energy prices, the scenarios can be used for analysis of the cost performance of new technologies and to serve as a basis for analysis of rates of market penetration. Differences in the scenarios among the three modeling groups give an impression of the types of market challenges that new options will face.

In addition, these studies could form the foundation of analysis of the non-climate environmental implications of implementing potential new energy sources at a large scale. Such analysis was beyond the scope of the present research, but information is provided that could form a basis for such analysis, for example, the potential effects on the U.S. and the globe of implied volumes of CCS and biomass production or of nuclear power expansion in some of the scenarios.

The scenarios could also be used in comparative mode. Just as many lessons were learned by comparing the differences between the three modeling groups' scenarios, still more could be learned by extending the comparison to scenarios that predate these or come after, including scenarios developed using entirely different approaches. For example, some scenario exercises do not apply economic models with detailed analysis of energy markets of the type used here. Such scenarios could be compared against those presented here to gain insight into the role of economic factors.

Finally, these scenarios might be used to explore the economic effects of stabilization at different levels. Such work was beyond the scope of the research specified in the Prospectus. However, the scenarios do contain information that can be used to calculate indicators of consumer impact in the U.S., for example, by using the changes in prices and quantities of fuels in moving from one stabilization level to another. (The reader is reminded, however, that these welfare effects do not include the benefits that alternative stabilization levels might yield in reduced climate change risk or ancillary effects, such as effects on air pollution).

MOVING FORWARD

As noted earlier, this work is neither the first nor is it likely to be the last of its kind. Throughout the report, a number of limitations to the approach and the participating models have been highlighted. Studies such as the one presented here would benefit from further research and model development and this section suggests several productive paths to pursue.

Technology Sensitivity Analysis

The importance of future technology development is clear in this report, and sensitivity testing of key assumptions would be of use. For example, what are the implications of various non-climate constraints on nuclear power or on the large-scale expansion of CCS or biofuels production? If particular supply technologies – nuclear, wind, natural gas combined cycle generation, and biomass – were assumed to be more or less expensive, how would that affect market penetration and policy cost? On the demand side, what are the effects of alternative views of the technical developments needed to facilitate substitution of electricity for liquid and gaseous fuels in various sectors, particularly in transport? Since technology deployment will be influenced by the policy environment, how would the consideration of less optimistic policy regimes affect this aspect of the scenarios?

Consideration of Less Optimistic Policy Regimes

The discussion in Chapter 4 emphasizes that the difficulty of the stabilization task emerging from any scenario research is crucially dependent on underlying institutional assumptions, and the insight to be gained from a single representation of control policy such as the one adopted in this research is limited. The scenarios assume a wide array of idealized institutions both in individual nations and in the international community. Both developed and developing economies are assumed to possess markets that efficiently pass price information to decision makers. Rules and regulations ranging from accounting and property rights to legal and enforcement systems are assumed to operate efficiently. While such assumptions provide a well-defined reference scenario and lower-bound information on potential costs, the probability is low that the world will actually implement such an idealized architecture. In that light, a natural direction for future research is to supplement the analysis presented here with analyses of policy regimes that are under discussion by nations and international organizations and that have a greater potential for being implemented. Such research would broaden the understanding of the stabilization challenge in areas ranging from technology development to the economics of global mitigation.

Expansion and/or Improvement of the Land-Use Components of the Models

A significant weakness in this research is the handling of the role of forest and agricultural sinks and sources. The major reason for this gap is that the models employed here were not well suited to analyze some of the complexities of this aspect of the carbon cycle. Yet, as this analysis has shown, agriculture, land-use and terrestrial carbon cycle issues play an important role in shaping the long-term radiative character of the atmosphere. Research that would improve the characterization of land use and land cover as well as improve the linkages among energy and economic systems, land use, land cover, terrestrial carbon processes, and other bio-geo-chemical cycles has potentially high payoff.

Inclusion of other Radiatively Important Substances

The focus in this research is on the relatively long-lived GHGs, but shorter-lived substances, such as ozone and aerosols, have strong radiative effects as well. More complete analysis would include these short-lived contributors, and their control possibilities, directly within the scenario analysis.

Decision Making under Uncertainty

Finally, the problem of how to respond to the threat of climate change is ultimately a problem of decision making under uncertainty that requires an assessment of the risks of climate change and how policies might reduce the odds of extremely bad outcomes. One would like to compare the expected benefits of policies to reduce GHG emissions against the expected costs of achieving those reductions. By focusing only on emission paths that would lead to stabilization, this research considers only the costs of stabilization without consideration of the benefits. Moreover, given the direction provided in the Prospectus, this research focused on scenarios and not on uncertainty analysis. It is not possible to attach probabilities to scenarios constructed in this way; formal probabilities can only be attached to a range, which requires exploration of the effects of many uncertain model parameters.

APPENDIX A

Accompanying this report are two databases of scenario information collected from the participating modeling teams. These databases can be found at http://www.climatescience.gov/.

The database entitled *CCSP 2_1A Scenario Information.xls* includes all the information that was collected from the modeling teams to support the development of this report, including the figures and tables presented in the report. The database entitled *CCSP 2_1A Other Species.xls* includes additional emissions information for substances not included in the radiative forcing stabilization levels used in this research, such as ozone precursors and aerosols. The emissions profiles for these other substances are different from what would emerge in scenarios in which these substances are explicitly included in stabilizing radiative forcing.

The two data files include the following fields:

Region
Category
Sub Category
Variable
Units
Run Label
Year

ACRONYMS

AEEI	autonomous energy efficiency improvement
AOGCMs	atmosphere-ocean general circulation models
CCS	carbon capture and storage
CCSP	Climate Change Science Program
CCTP	Climate Change Technology Program
CFCs	chlorofluorocarbons
CGE	computable general equilibrium
CPDAC	Climate Change Science Program Synthesis and Assessment Product
DOE	U.S. Department of Energy
EMF	Energy Modeling Forum
EPPA	Emissions Prediction and Policy Analysis
FCCC	U.N. Framework Convention on Climate Change
GCMs	general circulation models
GDP	gross domestic product
GHGs	greenhouse gases
GWP	Global Warming Potential
HFCs	hydrofluorocarbons
HCFC	hydrochlorofluorocarbons
IAMs	Integrated Assessment Models
IGCC	Integrated Gasification Combined Cycle
IGSM	Integrated Global Systems Model
IPCC	Intergovernmental Panel on Climate Change
MAC	marginal abatement cost
MAGICC	Model for the Assessment of Greenhouse-Gas Induced Climate Change
MER	market exchange rate
MERGE	Model for Evaluating the Regional and Global Effects
NCAR	National Center for Atmospheric Research
NGCC	natural gas combined cycle
NMVOCs	non-methane volatile organic compounds
OECD	Organization for Economic Cooperation and Development
PFCs	perfluorocarbons
PPP	purchasing power parity
SRES	Special Report on Emissions Scenarios
TAR	Third Assessment Report
U.N.	United Nations
U.S.	United States

Units

$2000	U.S. 2000 dollars
bbl	barrel
c/kWh	cents per kilowatt hour
EJ	exajoule
gal	gallon
GJ	gigajoule
Gt	gigatonne
GtC	gigatonne carbon
Mt	megatonne
MtC	megatonne carbon
Kt	kilotonne
PgC	petagram carbon
ppbv	parts per billion by volume
ppmv	parts per million by volume
ppt	parts per trillion
Quad	quadrillion btu
tcf	thousand cubic feet
W/m^2	watts per meter squared
yr	year

Chemical Formulas

CH_4	methane
CO	carbon monoxide
CO_2	carbon dioxide
N_2O	nitrous oxide
NO_x	nitrogen oxides
O_3	ozone
SF_6	sulfur hexafluoride

Babiker M., M. Bautista, H. Jacoby, and J. Reilly, 2000: *Effects of Differentiating Climate Policy by Sector: A U.S. Example*. Report No. 73, MIT Joint Program on the Science and Policy of Global Change.

Babiker M, G. Metcalf, and J. Reilly, 2003a: Tax distortions and global climate policy. *Journal of Economic and Environmental Management* **46**:269-287

Babiker M, L. Viguier, J. Reilly, A. Ellerman, and P Criqui, 2003b: The welfare costs of hybrid carbon policies in the european union. *Environmental Modeling and Assessment* **8**:187-1

Babiker M., J. Reilly, and L Viguier, 2004: Is emissions trading always beneficial. *The Energy Journal* **25**(2):33-56.

Brenkert A, S. Smith, S. Kim, and H. Pitcher, 2003: *Model Documentation for the MiniCAM*. PNNL-14337, Pacific Northwest National Laboratory, Richland, Washington.

CCSP, 2003: *Strategic Plan for the U.S. Climate Change Science Program*. A Report by the U.S. Climate Change Science Program and the Subcommittee on Global Change Research. Washington, D.C.

CCSP, 2005: *Final Prospectus for Synthesis and Assessment Product 2.1*. A Report by the U.S. Climate Change Science Program and the Subcommittee on Global Change Research. Washington, D.C.

CCSP, 2007: *Global Change Scenarios: Their Development and Use*. A Report by the U.S. Climate Change Science Program and the Subcommittee on Global Change Research [Parson, E., Burkett, V., Fisher-Vanden, K., Keith, D., Mearns, L., Pitcher, H., Rosenzweig, C., and Webster, M.]. U.S. Department of Energy, Washington, D.C.

CEA, 2005: *Economic Report of the President*, Council of Economic Advisors .U.S. Government Printing Office, Washington, D.C.

Clarke L., J. Edmonds, S. Kim, J. Lurz, H. Pitcher, S. Smith, M. Wise, *2007: Documentation for the MiniCAM CCSP Scenarios*. Battelle Pacific Northwest Division Technical Report, PNNL-16735.

Claussen M., L. Mysak, A.Weaver, M Crucifix, T. Fichefet, M. Loutre, S. Weber, J. Alcamo, V. Alexeev, A. Berger, R. Calov, A. Ganopolski, H. Goosse, G. Lohman, F. Lunkeit,

I.. Mokhov, V. Petoukhov, P. Stone, and Z. Wang, 2002: Earth System Models of Intermediate Complexity: Closing the Gap in the Spectrum of Climate System Models. *Climate Dynamics* 15:579-586.

Cubasch U, G. Meehl, G. Boer, R. Stouffer, M. Dix, A. Noda, C. Senior, S. Raper, K. Yap, 2001: 2001: Projections of Future Climate Change. In IPCC, 2001: *Climate Change 2001: The Scientific Basis*, 525-582. Cambridge, U.K.

de la Chesnaye, F. and Weyant, J (eds.), 2006: Multi-greenhouse gas mitigation and climate policy. *The Energy Journal*, Special Issue.

Edmonds, J., P. Freund, and J. Dooley, 2001: The Role of Carbon Management Technologies in Addressing Atmospheric Stabilization of Greenhouse Gases. In: *Greenhouse Gas Control Technologies, Proceedings of the Fifth International Conference on Greenhouse Gas Control Technologies*, D.J. Williams, R.A. Durie, P. McMullan, C.A.J. Paulson and A.Y. Smith (eds.) CSIRO, Collingswood, VIC, Australia. pp:46-51.

DOE, 1985: *Atmospheric Carbon Dioxide and the Global Carbon Cycle*, ed JR Trabalka, DOE/ER-0239. Office of Energy Research, United States Department of Energy, Washington, D.C.

DOE, 2006: *Short-Term Energy, and Winter Fuels Outlook*, October 10th, 2006 Release. United States Department of Energy, Washington, D.C.

Dooley J. and Friedman S., 2004: *A Regionally disaggregated global accounting of CO_2 storage capacity: data and assumptions*. Battelle Pacific Northwest Division Technical Report Number PNWD-3431.

Edmonds J. and J Reilly, 1985: *Global Energy: Assessing the Future*. Oxford University Press, New York.

Hotelling H., 1931: The economics of exhaustible resources. *Journal of Political Economy* 39:137-175.

IPCC, 1991: *Climate Change: The IPCC Response Strategies*. Intergovernmental Panel on Climate Change, Island Press, Washington, D.C.

IPCC, 1992: *Climate Change 1992: The Supplementary Report to the IPCC Scientific Assessment*. eds JT Houghton, BA Callander, and SK Varney, Intergovernmental Panel on Climate Change, Cambridge University Press, Cambridge, U.K.

REFERENCES

IPCC, 1996: *Climate Change 1995: Economic and Social Dimensions of Climate Change.* eds JP Bruce, H Lee, and EF Haites, Intergovernmental Panel on Climate Change, Cambridge University Press, Cambridge, U.K.

IPCC, 2001: *Climate Change 2001: The Scientific Basis. The Contribution of Working Group I to the Third Assessment Report of the Intergovernmental Panel on Climate Change*, 944. eds JT Houghton, Y Ding, DJ Griggs, M Noguer, P. van der Linden and D. Xiaosu, Intergovernmental Panel on Climate Change, Cambridge University Press, Cambridge, U.K.

Leggett J, W. Pepper, R. Swart, J Edmonds, L. Meira Filho, I. Mintzer, M. Wang, and J. Wasson, 1992: Emissions Scenarios for the IPCC: An Update. In: *Climate Change 1992: The Supplementary Report to the IPCC Scientific Assessment*, University Press, Cambridge, U.K.

Maddison, A., 2001: *The World Economy: A Millennial Perspective.* Organization for Economic Cooperation and Development, Paris, France.

Maier-Reimer E. and K .Hasselmann, 1987: Transport and Storage of CO_2 in the Ocean —An Inorganic Ocean-Circulation Carbon Cycle Model. *Climate Dynamics* 2:63–90.

Manne A and R Richels. 2005. MERGE-A Model for Global Climate Change. In: *Energy and Environment*, eds R. Loulou, J. Waaub, and G. Zaccour, Springer, New York.

Manne A. and R. Richels, 1994: The costs of stabilizing global CO_2 emissions: a probabilistic analysis based on expert judgements. *The Energy Journal.* 15(1):31-56.

Manne A. and R. Richels, 2001: An alternative approach to establishing tradeoffs among gases. *Nature* 419:675-676.

Morita T. and H-C. Lee, 1998: IPCC SRES Database, Version 0.1, Emission Scenario Database Prepared for IPCC Special Report on Emission Scenarios. Accessed at http://www cger.nies.go.jp/cger e/db/ipcc.html.

Nakicenovic N., J. Alcamo, G. Davis, B. de Vries, J. Fenhann, S. Gan, K. Gregory, A. Grubler, T. Jung, T. Kram, E. Rovere, L. Michaelis, S. Mori, T. Morita, W. Pepper, H. Pitcher, L. Price, K. Riahi, A. Roehrl, H. Rogner, A. Sankovski, M. Schlesinger, P. Shukla, S. Smith, R. Swart, S. van Rooijen, N. Victor, and Z. Dadi, 2000: *Special Report on Emissions Scenarios*. Cambridge University Press, Cambridge, U.K.

Nakicenovic N., P. Kolp, K. Riahi, M. Kainuma, T. Hanaoka, 2006: Assessment of Emissions Scenarios Revisited. *Environmental Economics and Policy Studies*, 7(3):137-173

Nakicenovic N, N Victor, and T Morita, 1998: Emissions scenarios database and review of scenarios. *Mitigation and Adaptation Strategies for Global Change* **3(2–4)**:95–120.

NAS, 1983: *Changing Climate: Report of the Carbon Dioxide Assessment Committee.* - National Academy of Sciences, National Academy Press, Washington, D.C.

NOAA - National Oceanic and Atmospheric Administration, 2007: Global Warming: Frequently Asked Questions, at http://lwf.ncdc. noaa.gov/oa/climate/globalwarming.html.

Nordhaus W. and G. Yohe, 1983: Future Carbon Dioxide Emissions from Fossil Fuels. In: **Changing Climate**, 87-153. National Academy Press, Washington, D.C.

NRC, 2005: *Radiative Forcing of Climate Change: Expanding the Concept and Addressing Uncertainties.* National Research Council, National Academy Press, Washington, D.C.

O'Neill B., 2005: Population Scenarios Based on Probabilistic Projections: An Application for the Millennium Ecosystem Assessment. *Population and Environment* **26(3)**:229-254.

Paltsev S., H. Jacoby, J. Reilly, L. Viguier and M. Babiker, 2005: Modeling the Transport Sector: The Role of Existing Fuel Taxes. *Energy and Environment*, pp.211-238 eds R. Loulou, J-P. Waaub, and G. Zaccour, Springer, New York.

Paltsev S., J. Reilly, H. Jacoby, R. Eckaus, J. McFarland, M. Sarofim, M. Asadoorian, and M. Babiker, 2005: *The MIT Emissions Prediction and Policy Analysis (EPPA) Model: Version 4*. Report 125, MIT Joint Program on the Science and Policy of Global Change, Cambridge, Massachusetts.

Parson E. and K. Fisher-Vanden, 1997: integrated assessment models of global climate change. *Annual Review of Energy and the Environment* 22:589-628.

Peck S. and Y. Wan, 1996: Analytic Solutions of Simple Greenhouse Gas Emission Models. Chapter 6 in: *Economics of Atmospheric Pollution*, eds EC Van Ierland and K Gorka, Springer Verlag, New York.

Pepper W., J. Leggett, R. Swart, J. Wasson, J. Edmonds, and I. Mintzer, 1992: Emissions Scenarios for the IPCC: An Update – Assumptions, Methodology, and Results. Intergovernmental Panel on Climate Change, Geneva, Switzerland.

Prinn R,.,J. Reilly, M. Sarofim, C. Wang and B. Felzer. In Press. Effects of air pollution control on climate. In: *Integrated Assessment of Human-induced Climate Change*, ed M Schlesinger, Cambridge University Press, Cambridge, U.K.

Raper, S., J. Gregory, and T. Osborn, 2001: Use of an upwelling-diffusion energy balance climate model to simulate and diagnose a/ogcm results. *Climate Dynamics* **17**:601-613.

Reilly J., J. Edmonds, R. Gardner, and A. Brenkert, 1987: Uncertainty analysis of the aea/orau CO_2 emissions model. *The Energy Journal* **8(3)**:1-29.

Reilly J., H. Jacoby, and R. Prinn, 2003: *Multi-Gas Contributors to Global Climate Change: Climate Impacts and Mitigation Costs of Non-CO₂ Gases*. Pew Center on Global Climate Change, Washington, D.C.

Richels R. and J. Edmonds, 1995: The economics of stabilizing atmospheric CO_2 concentrations. *Energy Policy* **23(4/5)**:373-78.

Richels R., J. Edmonds, H. Gruenspecht, and T. Wigley, 1996: The Berlin Mandate: The Design of Cost-Effective Mitigation Strategies. *Climate Change: Integrating Science, Economics and Policy* CP-96-1:29-48. Nakiccnovic N, WD Nordhaus, R Richels, and FL Toth (eds.), International Institute for Applied Systems Analysis, Laxenburg, Austria.

Sabine C., M Heiman, P. Artaxo, D. Bakker, C. Chen, C. Field, N. Gruber, C. LeQuéré, R. Prinn, J. Richey, P. Romero-Lankao, J. Sathaye, and R. Valentini, 2004: Current Status and Past Trends of the Carbon Cycle. Pages 17-44 in *The Global Carbon Cycle, SCOPE Project 62*, eds CB Field and MR Raupach, Island Press, Washington D.C.

Sands, R., and M. Leimbach, 2003: Modeling agriculture and land use in an integrated assessment framework, *Climatic Change* **56(1)**:185-210.

Sarofim M., C. Forest, D. Reiner, and J. Reilly, 2005: Stabilization and Global Climate Policy, *Global and Planetary Change* 47:266-272. Accessed as MIT Joint Program on the Science and Policy of Global Change Reprint No. 2005-5 at http://mit.edu/globalchange/www/.

Scott M., R. Sands, J. Edmonds, A. Liebetrau, and D. Engel, 2000: Uncertainty in integrated assessment models: modeling with minicam 1.0. *Energy Policy* **27(14)**:855-879.

Smith, S. and Wigley, T., 2006: Multi-Gas Forcing Stabilization with Minicam, *The Energy Journal*, Multi-Greenhouse Gas Mitigation and Climate Policy Special Issue.

Smith S., 2005: Income and pollutant emissions in the objects minicam model. *Journal of Environment and Development* **14(1)**:175–196.

Sokolov A., C. Schlosser, S. Dutkiewicz, S. Paltsev, D. Kicklighter, H. Jacoby, R. Prinn, C. Forest, J. Reilly, C. Wang, B. Felzer, M. Sarofim, J. Scott, P. Stone, J. Melillo, and J. Cohen, 2005: *The MIT Integrated Global System Model (IGSM) Version 2*: Model Description and Baseline Evaluation. Report 124, MIT Joint Program on the Science and Policy of Global Change, Cambridge, Massachusetts.

UN, 1992: *Framework Convention on Climate Change*. United Nations, New York.

UN, 2000: *Long-Run World Population Projections: Based on the 1998 Revision*. United Nations, New York.

UN, 2001: *World Population Prospects: The 2000 Revision, Data in digital form*. Population Division, Department of Economic and Social Affairs, United Nations, New York.

UN, 2005: *World Population Prospects: The 2004 Revision, Data in digital form*. Population Division, Department of Economic and Social Affairs, United Nations, New York.

Webster M., M. Babiker, M. Mayer, J. Reilly, J. Harnisch, R. Hyman, M. Sarofim, and C. Wang, 2003: Uncertainty in emissions projections for climate models. *Atmospheric Environment*, **36(22)**:3659-3670.

Webster M., M. Babiker, M. Mayer, J. Reilly, J. Harnisch, R. Hyman, M. Sarofim, and C. Wang, 2002: Uncertainty in emissions projections for climate models. *Atmospheric Environment*, **36(22)**:3659-3670.

Webster M., 2003: Communicating climate change uncertainty to policy-makers and the public. *Climatic Change* **61**:1-8.

Wigley T. and S. Raper, 2001: Interpretation of high projections for global-mean warming. *Science* **293**:451-454.

Wigley T. and S. Raper, 2002: Reasons for larger warming projections in the ipcc third assessment report. *Journal of Climate*, **15**:2945-2952.

Wigley T., R. Richels, and J. Edmonds, 1996: Economic and environmental choices in the stabilization of atmospheric CO_2 concentrations. *Nature*. **379(6562)**:240-243.

Wigley, T., Smith, S., Prather, M., 2002: Radiative forcing due to reactive gas emissions, *Journal of Climate* **15(18)**, pp. 2690-2696.

CONTACT INFORMATION

Global Change Research Information Office
c/o Climate Change Science Program Office
1717 Pennsylvania Avenue, NW
Suite 250
Washington, DC 20006
202-223-6262 (voice)
202-223-3065 (fax)

The Climate Change Science Program incorporates the U.S. Global Change Research Program and the Climate Change Research Initiative.

To obtain a copy of this document, place an order at the Global Change Research Information Office (GCRIO) web site: http://www.gcrio.org/orders

CLIMATE CHANGE SCIENCE PROGRAM AND THE SUBCOMMITTEE ON GLOBAL CHANGE RESEARCH

William J. Brennan, Chair
Department of Commerce
National Oceanic and Atmospheric Administration
Acting Director, Climate Change Science Program

Jack Kaye, Vice Chair
National Aeronautics and Space Administration

Allen Dearry
Department of Health and Human Services

Jerry Elwood
Department of Energy

Mary Glackin
National Oceanic and Atmospheric Administration

Patricia Gruber
Department of Defense

William Hohenstein
Department of Agriculture

Linda Lawson
Department of Transportation

Mark Myers
U.S. Geological Survey

Jarvis Moyers
National Science Foundation

Patrick Neale
Smithsonian Institution

Jacqueline Schafer
U.S. Agency for International Development

Joel Scheraga
Environmental Protection Agency

Harlan Watson
Department of State

EXECUTIVE OFFICE AND OTHER LIAISONS

Melissa Brandt
Office of Management and Budget

Stephen Eule
Department of Energy
Director, Climate Change Technology Program

Katharine Gebbie
National Institute of Standards & Technology

Margaret McCalla
Office of the Federal Coordinator for Meteorology

George Banks
Council on Environmental Quality

Gene Whitney
Office of Science and Technology Policy